PIG PRODUCTION

What the textbooks don't tell you

John Gadd

Nottingham
University Press

First published by Nottingham University Press

This reissued original edition published 2023 by 5m Books Ltd
www.5mbooks.com

British   Library   Cataloguing   in   Publication   Data
Pig Production: What the Textbooks Don't Tell You

ISBN 9781789182996

Disclaimer

Typeset by Nottingham University Press, Nottingham

EU GPSR Authorised Representative
LOGOS EUROPE, 9 rue Nicolas Poussin, 17000, LA ROCHELLE, France
E-mail: Contact@logoseurope.eu

CONTENTS

ACKNOWLEDGEMENTS

'No man is an island...', so John Donne said. And neither are technical writers like myself.

All of us build on the experience of others, and the subjects covered in this book start from the findings of a wide number of authorities on pig production and the science behind them. I have tried to add some comments of my own from 40 years' experience at the sharp end of giving on-farm advice.

In particular I need to acknowledge my debt to the following pig experts, all of whom have influenced my thinking. These are in no order of preference, but as I am reminded of them when reading the final drafts.

Prof Peter Brooks and his team at the former Seale-Hayne College (now part of Plymouth University) for their valuable pioneer work on the no-longer 'forgotten nutrient' – water. *Mark Blackwell* of Antec International and their pig adviser, veterinarian *Jake Waddilove* for their good sense on biosecurity, especially modern sanitation methods and the correct use of new uprated products. *Dr Paul Hughes* for his pioneering work on breeding gilts and re-breeding sows. *Dr Pearse Lyons*, CEO of the nutritional biotechnology pioneers Alltech Inc USA, for his inspiration and friendly guidance. *Dr Gordon Rosen*, Nutrition Consultant, for being a patient sounding-board in the field of correct scientific terminology and its expression. Veterinarian *Ray Blackburn* for his advice on splay-legs. *Dr François Madec* for his eminent foresight and practical comments on how to mitigate the scourge of PMWS. Pig veterinarian *Neville Kingston* for his always-sensible advice on repopulation techniques. And his sadly-missed colleague, the late *Michael Muirhead*, so approachable and generous with his advice, too.

My ex-colleagues in the animal feed trade, *Paul Toplis* and *Mick Hazzledine* for keeping me up-to-date in this field. *Prof Colin Whittemore*, of Edinburgh University, for pre-eminent commonsense in so many pig areas. Two more eminent academics, also in Scotland, *Drs Peter English* and *Vernon Fowler*, both crystal-clear thinkers. Pioneer baby-pig nutrition specialist *Dr Mike Varley*, SCA Ltd, for advice and encouragement on getting to grips with this vital subject. Staff at the now discontinued *Pig Demonstration Unit* at Stoneleigh for the stomach-tubing photographs. And this establishment fortunately now replaced by *MLC Stotfold Pig Experimental Farm* headed by the inspirational *Dr Pinder Gill* and his team – always ready to answer my tiresome questions.

Dr Temple Grandin (Colorado USA) for her original thinking on pig movement and handling. American research veterinarians led by *Dr Tim Stahly*, of Iowa State University USA and supported by *Dr Ian Williams* in Australia, for their good lead on how immune status affects nutrient demand. The late *David Taylor MBE* for his occasionally infuriating but very down-to-earth advice when I was young on how to make a profit out of pig farming. Salesmen extraordinaire *Reg Hardy* (UK), *Mike King* (USA) and the late *Norman Tucker* (UK) who by their incomparable example have taught me so much about successful selling to pig farmers. *Drs Des Cole* and *William Close* for concise and well-written advice in their excellent recent book 'The Nutrition of Sows and Boars'.

To the staff at NUP headed by the same Dr Cole, and especially *Sarah Keeling*, who have designed and issued this book. Likewise to my long-suffering secretary, *Janeen Evans*, who patiently retyped the many alterations I've made to what she must have prayed were final, finished drafts!

And to *Barbara, my wife* of 50 golden years now, who holds the fort while I am away for long periods gathering material, or when shut away in the office writing it all up into article or book format.

To the *30 or more pig farmers* who have, over many years, helped with farm trial work so as to shed more light into the darker corners of pig farming. Too many to mention, but they should know that I am sure that they have materially contributed to the knowledge of pig production.

I also thank magazine publishers *RBI** for generously granting permission for NUP to base this book on some of my articles written for the global magazine 'Pigs' (now 'Pig Progress') and *Anabel Evans*, the current editor, for making the vital introductions for me.

Finally, I may have left out acknowledging help and advice from some individuals and companies which my faulty 35-year memory could have inadvertently excluded. My apologies to you, and if you wish to contact me through the publishers for such omissions to be rectified in any later edition, please do so.

---

*Full title: Reed Business Information (International Agriculture and Horticulture)

INTRODUCTION

I have about 35 textbooks on pig production in my pig library. And very worthwhile volumes they are. Over the years I've consulted them frequently to check on facts and to remind myself of the hundreds of pieces of information which my fickle memory has let slip across 40 years of advisory work in and around pigs.

What tremendous knowledge, wisdom, experience, hard work, valid opinion and comment they contain! Some textbooks are old friends, well-thumbed after many a "What does So-and-So think about it" consultation. Some have tended to gather dust – not their fault, but mine, as they were rather too specialized and detailed for my limited intelligence, but I felt I ought to buy them to be on the safe side.

About 20 years ago I noticed when I looked for the fact or reference I needed, that rather too often I found myself thinking, 'Yes, but…' or 'But what if…?' or 'He hasn't indicated the likely cost…' or 'What was the disease/environmental situation,' or even 'Good Heavens, he seems to have left that bit out altogether!'.

Presumptuous

I realise it is certainly presumptuous of me, and possibly impertinent, to think like that about the work of other authors. After all, their textbooks show them to be far superior to me in their breadth of knowledge in their particular subject. Nevertheless these thoughts did occur often enough for me to try out the idea of maybe writing a few pieces on what these textbooks didn't cover (of course mentioning no names, or book titles) and submitting them to the editors of two pig magazines to see what they thought about a series.

One editor declined – I learned later that he didn't trust me not to drop a hint or two as to which textbook I might be – he thought – criticising. But the other editor, Wiebe van de Sluis of the magazine 'Pigs' (now 'Pig Progress') courageously took up the idea and let me put in two or three pieces. He liked them, and went on to publish them. That was 15 years ago, and the series is still in being as I write under another editor, Anabel Evans' watchful eye, with over 150 subjects covered. So many in fact, that pig farmers all over the world have said to me, "Why don't you publish them in book form – I've kept a few, but must have missed many."

Generosity

I raised this idea with the publishers and copyright owners, Reed Business Information, who are mainly specialist magazine publishers, and they said a book is not really their scene. But they generously allowed me to pass the idea across to Nottingham University Press (NUP) who have issued this modified and completely updated collection of some of the typical, and I hope most useful, subjects I wrote about over those 15 years. To bring things up-to-date I've added a few more in passing.

So, thank you, RBI. This is an unusual, probably exceptional, act of generosity in the competitive publishing world, and both Nottingham University Press and I are very grateful for the magnanimity you have shewn.

No criticism implied

As I said, the first editor I contacted was worried about my criticising specific textbooks or their authors. This has never been my intention and never will be. This book reveals some of the hidden corners – what the French call 'les introuvables' (the 'unfindables') – in the vast subject of pig production, which have not yet been dealt with by many textbooks, or (in my opinion) not adequately covered. These omissions must be mainly due to lack of space, I'm sure, as most textbook authors will know much more about their subject than I do, and are perfectly capable of filling in the gaps.

The biter bitten!

Last year, NUP published my own textbook, all 590 pages of it, on 'Pig Production Problems' and my suggested solutions to them. I am happy to confess (and quite amused by it as well) that readers have drawn my attention to quite a few points that my *own* textbook didn't tell you about on the subjects I covered, so I have included them here. Thus the biter has been well and truly bitten by his own book.

And that – if not poetic– is certainly literary justice!

Some suggestions

While I'm on the subject of rocking the textbook boat, so to speak, here are three things I would like to see in future textbooks on pigs. Again, presumptuous of me to say so, but this is the right place to say it! So here goes.

Authors . . .

1.   When providing a table, putting a sentence or two under it saying what the figures mean, or could suggest, is very helpful to us laymen. After all, for whom are you writing the textbook? Other scientists who don't need the explanation, or for students and the farming public, who do?

2.   Wherever possible please try and flesh out the positive or negative result of a trial with some econometric figures – 'what this might mean in dollars and cents' as the Americans say.

Many scientists are aghast at this radical suggestion. But as long as the author *sets out clearly the economic assumptions he has made in coming to his conclusion, it does not matter if the economics or currency factors are completely different between pig industries or individual farms*. As long as the researcher provides a matrix (framework) of how he calculated his econometrics, the reader can replace the researcher's figures with those of his own in order to see how the trial result might affect himself.

This is good communication. Too many scientists are not good at communicating, are they? Only to themselves!

The academics I've spoken to reject the suggestion on two grounds: "It is not science"; and "I am too busy to go on to do it anyway".

Sure, it is not science, but it is, to quote the TV ad, 'The Appliance of Science'. When I've done these exercises, sometimes the statistically beneficial research result was, at today's costings, too expensive to make it worthwhile. Conversely some negative physical benefits (on creep feed or end-of-nursery pigs for example) when extrapolated to slaughter weight, nevertheless could have secured a comfortable extra profit! That's when we sell the pigs, not at weaning or (rarely) at the end of the nursery stage.

Too busy to do it? Actually, when you have done a few as I have, the process is quite quick, especially if you create your own matrix. And, at the end of the day, research salaries are paid for by the farmer through the taxpayer! Progressive pig farmers are becoming more hard-nosed and familiar with the need for a bottom line (or brought up short by the lack of it) and worthy physical, statistically-accurate research which has gone that extra mile to express the results in an economic 'what if' form will be welcome – and ipso facto, the careers of academics safeguarded in the long run, too.

"No, *you* farm management people can do it for us," say some of my research friends, "and leave us to get on with the science." Sure (I try to do it in this book), but it looks better coming from you. After all, it is your work, not mine and you should get the extra credit for it.

I commend the pragmatic – and successful – Danes as the first research trialists to adopt the idea as general routine. Many of the Danish Pig Research Institute's trials carry an economic assessment (in Kroner - Euros as well would be more helpful, my Danish friends!).

3. One last point. Less scientific jargon would be nice! Follow the examples of highly-respected scientists like Colin Whittemore, John McGlone, Frank Aherne, Peter Brooks, Bill Close, Vernon Fowler and veterinarian John Carr, whose papers and articles are models of clarity. O rarae aves!

So. . . a 'dip-into' book

Because I've kept the subjects short, concentrating on their practical and cost-effective points, this is very much a 'dip-into' book. I've always liked dip-into technical books if they are laid-out well with a decent index, and that is what we have tried to do here. Technical subjects can be heavy going, so I've tried to lighten the load for you as much as I can without, I hope, jeopardising the impact or the accuracy.

Enjoy this collection of hidden corners in our worldwide pig industry.

Running an efficient pig business

HOW TO BUY PIGFEED

It is perhaps worthwhile to commence this series of subjects with the first one submitted to 'Pigs' (as the current magazine 'Pig Progress' was called in 1989). This was – and still is – the very topical one of 'How to Buy Feed'. I reproduce the original text below, containing as it does my advice of 16 years ago. *However, only question 1 and the last two paragraphs I would retain today*!

Even so, this advice was based on some valuable sharp-end experience. I had been involved in marketing mineral and vitamin supplements with a national company for 5 years. Then had 6 years as a buyer (among other jobs) for a feed supplement company and for their 1200 sow farm. Then as pig product manager for Britain's second largest animal feed manufacturers for 12 years.

Further on, on page 8. I give a fully updated version to meet today's conditions.

What I said in 1989

Note: This is advice given 16 years ago, which has subsequently been superseded. It is quoted as an interesting comparison to how things have changed in this particular area of pig nutrition. I wrote in 1989...

"I note with amusement that half my working life has been spent persuading people to buy feed. Now I'm on my own, an equal period has been spent advising farmers to check up on the approaches made to them by the feed company reps!

I find many pig producers buy feed in a very unstructured way. On price; on protein levels; on energy estimation; on a whim – usually a combination of all four. Feed is your commonest and most profit-affecting purchase so it pays to adopt a more rational approach.

I append a list of questions you ought to ask to sort out the best likely buy. It shows you which is sensibly-costed 'cheap' food, compared to over-priced excellence!

3

If you feel it is all too advanced for you, then get someone who is independent to help you with the replies you get - that's what advisers are for. After a while you'll soon get the hang of it.

*Even then, at the end of the day you need to test out your two top 'best buys' by having designed for you a statistically-valid farm trial, carrying it out diligently, then getting the results interpreted by an adviser with statistical training to tell you what you can and **can't** say about them.*

So you're going to need help anyway. Next month I'll tell you what you need to do to get a simple farm trial mounted correctly. In the meantime, here is the first step.

1.   *Question:*

"You know my pigs, my housing and management. You know your feeds – I don't. Please recommend a feeding programme on your feeds which you are confident will give me the greatest profit, including grading, growth and FCR."

Reason for asking it:

*It is essential to get commitment from the supplier. He should be designing his feeds, feed scale (and therefore price-list) to growth, FCR and grading targets, and you need to make him state these in respect to **your** standard of husbandry, not just anybody's.*

Comment: This question is fair enough for today's conditions too, and I'd still recommend approaching a feed manufacturer in this way. But from here on in what I suggest is a more pragmatic/realistic approach to choosing a feed (and feed supplier) which commences on page 5, the approach has been completely superseded.

I describe yesterday's advice which now follows to alert you to how different the approach is today.

So, to continue with yesterday's approach . . .

2.   *Question:*

"What sort of daily gain (g/day) will this programme be likely to achieve in 10 kg steps? Also the grading expectancy at the end of the feeding programme, what do you estimate this to be?"

Disclaimer:

*"I accept fully that no **guarantee** is expected from you on either of these estimates, but I want you to give me a reasonable figure, please, which you can justify."*

Reason for asking it:

*This ensures that he cannot just make bland assurances on other people's results or his own trial results (which may be 'weighted') and ties him down still further in respect to what he honestly thinks his feeds will do for your pigs under **your** conditions - nobody else's. You need this claim to compare cost/pig/produced. He **should** be able to give it. You MUST mention this disclaimer; if not, his legal adviser will rightly tell him not to go further!*

3.   Question:

"In the same 10 kg steps, what daily intake (g/day) of lysine do you expect? State whether this is total lysine, available lysine or digestible lysine."

Reason for asking it:

*This is the start of the information you need to compare one food technically in theory - with another. It is **not** a hard-and-fast factor, but it is far better than the uninformed guesswork pig producers use at present.*

4.   Question:

"In the same 10 kg steps, what daily intake of energy in MJ/day DE do you expect?"

Reason for asking it:

This is also essential information, similar to the above

5.   Question;

"What feed scale do you advise, also in 10 kg steps? kg/day, please. If ad-lib feeding is involved please give an assessment of what my pigs will eat under my conditions?"

Reason for asking it:

Now we have enough basic information to compare feeds, at least to sort the men out from the boys. Remember to compare like with like, i.e. total/available/ digestible lysine, ad-lib with ad-lib, restricted with restricted, etc.

6.   Question:

"What lysine level does your feed contain? Total, available or digestible? In % of feed, please."

Reason for asking it;

These two questions, 6 & 7, are needed to cross-check that the salesman is recommending the right nutrient density product to supply their stated needs in 3 & 4. They are 'out' rather too often for comfort; not intentionally necessarily, but due to human error somewhere along the line. It always pays you to do this crosscheck; it is **your** money!

7.   Question;

"What energy level(s) do(es) your feed(s) contain? In MJ/DE per kg, please."

Reason for asking it:

If the information in 7 & 5 doesn't marry up with 4; or 6 and 5 with 3; ask why

Up to here you have acquired from each supplier their estimate of what primary nutrients your pigs need per day to obtain the performance which you have made them declare their feeds should be capable of reaching under **your** particular conditions. Plot them out. They should be similar, but not identical. If one is way out, it could be an over-optimistic claim, or a badly-designed programme. Discard it.

Finally, and **only finally,** we need to determine which of these claimed performances (which, by-and-large, you will have to accept as valid) is the **cheapest** performance. So we need a quote.

8.   Question:

"What is your quotation please, at the trough, all discounts taken?"

It is now easy to establish a 'theoretical best-buy', because you have from each supplier: -

a)   **Daily gain estimate,** so you can work out how many days each supplier thinks your pigs will take to get to slaughter on his food.

b)   **Food consumed estimate** i.e. from his advised feed scale, so you can work out the amount of food, in kg, each pig will eat.

c)   **Cost/pig estimate**; just multiply the cost of the food in p/kg x no. of kg eaten. The lowest cost is the one to buy.

d)   Just in case this is at the expense of **grading** - in Q.2 you've asked him to nail his grading colours to the mast; so, as a last check, compare your 1st, 2nd, and 3rd **cost** choices with how each ranks in grading claims and do a further income-less cost sum on each. Take this into your final decision.

COMPLICATED

If you think this is too complicated, or unnecessary - with respect, you are wrong! A pig is what it eats and your livelihood largely depends on what food you pay for it to do this.

Feed manufacturers - you've had it too easy in the past! Away with platitudes and assurances, you are going to have to present a much more cost-effective case to more hard-nosed buyers in future.

On the whole the feed trade do an excellent job for us. But there is a good deal of difference between their feeds, company to company and even mill to mill within the same company. This article spills some beans in telling your customers how to spot the differences - which I find do exist. I'm sure the feed trade will rise to the occasion. They always have. Go to it, lads!

Why things are different today

While there is nothing radically wrong with what I suggested in those days, the knowledge of pig nutrition – especially that of the growing pig – has moved on to such an extent that the procedure outlined has been overtaken by feasibility. What I mean by this is that even if today's prospective buyer of the feed questioned the feed company as I outlined, and even if the majority of the answers were indeed secured, the end-result today might still not be best suited to the farm concerned.

Why? The big difference between the present day and 15 years ago is nutritionally we realise that individual farms are very different! Different environment (ventilation, insulation, stocking density, pen layout and shape), stockmanship, pathogen loadings, genetics (less variation than 15 years ago but still enough to explore the effects of different diets between breeds), appetite, docility/resistance to stressors and veterinary supervision/ monitoring. All these have an impact on how the pigs perform cost effectively – especially the growing pig which accounts for 75% of the feed cost of producing a kilogramme of saleable meat.

Even so, growing pig nutrition is still very much a developing science. I can already see on the horizon techniques and new means of delivering food to the pig's mouth which will help solve some of the complexities and make simpler, possibly even very simple, the ultimate goal of not only the 'feed for the breed' (which knowledge is here now), but also 'the feed for the environment/climate' (which is still emerging as pig farmers improve the way they house and manage the growing pig so as to stress it less). Finally, the ultimate ambition of the 'feed for the current immune status' which will be a major retaliatory defence against the drag of chronic disease[1].

So if I were to write this essay again with the advances – and the complexities – of 16 years technical progress in mind, what would I say today? I need to replace the 1989 text with solutions to a whole crop of

---

[1] You can read more about this in the sections on 'Immunity' and 'Modern Techniques . . .'

questions. These are.... How does the perplexed pig farmer deal with the complex variations involved in choosing the best feed? Which is the best feed firm to deal with? How can the farmer give himself the best chance of getting the right diet, at the right (not necessarily the lowest) price? How can he assess the value of one food on offer from another if used under his own conditions? Does he need to have a farm specific diet? If so, how?

So...here is a revised approach to: -

What the Textbooks Don't Tell You About...

...HOW TO BUY FEED

1.   Choose the right manufacturer

Easier said than done, I know! It is achieved these days by judicious questioning and patient research. Of course, at face value all the feed firms seem the same, large or small. However, with 21 years of private consultancy work, sitting in with pig farmers who needed help on looking into the results, good or bad – usually bad (otherwise I wouldn't have been called in) – of the feeds they were using, to add to my earlier 23 years experience of being on the feed manufacturer's side of the fence, all this encourages me to say the following:

Go beyond the feed salesman.

*   The feed rep is the manufacturer's front-man. Likeable and personable as he is, regard him or her solely as an intermediary. It is the buying, formulation and feed manufacturing departments who are critical to your choice of feed. The feed rep is a useful, even essential, go-between, but you need to dialogue at a higher level, too.

*   The key person to befriend up the line is the nutritionist/formulator. He/she has – or should have – influence on buying and manufacturing departments making, if possible, their jobs easier, and alternatively/ occasionally convincing them that the product design constraints he may have to put upon them are justified in terms of product performance and quality.

Why the nutritionist is a key person. You, as the customer who pays all the salaries of the head office staff he or she works with, must get to know the manufacturer's nutritionist. With the technical knowledge now available on the critical balance between pig-related net energy (not the ME or DE of 15 years ago) as well as ileal digestible amino acids (not the total or available lysine etc of 15 years ago) no farmer – even the younger ones with good degrees, can keep pace with the knowledge required to fine-tune the modern growing pig diet to fit the high performance pig digestive 'engine' of today.

A good nutritionist on top of his job will be able to design ('formulate' is the term they use) a diet so that it does this with the maximum of digestive efficiency. He uses the right raw materials bought-in to a known, or subsequently analysed nutritional specification. He must also make it fully 'manufacturable' to arrive fresh and nutritionally undamaged on your farm.

2.   **Your job is to choose the right food for your conditions – nobody else's.**

Your job is not to try to design your own food but to …

*   Dialogue with the nutritionist who designs your feed. This is done by providing him with as much information as possible about the genetics of your pigs, your market targets, your housing, the way you keep them, your current disease status and your current performance.

*   If the nutritionist, or his department/section is difficult to contact, or displays a lack of interest in these objectives then it is probably wise to try another company. In my experience of probing further this reluctant or difficult to contact attitude is usually because of pressure of work/inadequate investment in this, the powerhouse of the feed firm. Often this area seems to be undervalued by their accountants who control the money supply.

*   The better feed firms have accountants who *listen* to the nutritionists, the mill and transport managers, the buyers and the 'pbi' (the sales side) before making financial strictures.

*   *The commercial pig nutritionist*

     A good nutritionist displays interest in your farm, asks questions, answers your queries willingly, provides evidence, does what he says he will do promptly and briefs his salesmen well. He needs to be given the time and space to do this. Almost exclusively, in 21 years of dealing with company nutritionists from the sharp end and thorny side of the fence, a firm with such a paragon provides good food at a decent price, providing that it also has a good buying department. That is not to say that a firm with a more distant, aloof, feed design strategy doesn't, or wouldn't, or hasn't. But I know who I'd select as a first choice, all the same.

*   *The on-farm mixer*

     While this personal contact with the nutritionist is still advisable if you decide to buy complete feeds, it is absolutely vital if you are farm-mixing your own cereal/straights, and so needs to be moved up a plane. The variation in raw material analysis is enormous of the cereals you grow or the straights and co-products you can buy. No nutritionist worth half a packet of peas can formulate properly and use the gold-plated nutritional knowledge now available to him in the pig and poultry areas if he doesn't know the specification of the raw materials you will use. After all he has to spin them all together into a high-octane fuel for your Formula One pigs. Without this he has to guess at things – make assumptions, and that is not fine-tuned nutrition.

- ### *Recommendations from others*

   It used to be said that the producer should visit other producers on the same firm's food and discuss with them their satisfaction with your own preferred supplier. Well, maybe.

   In terms of results, i.e. pig performance (and even profit if your farmer contact will tell you), you need to consider their approval of the feed manufacturer they favour (or otherwise) guardedly, as their farms will be different to yours. What feed works for one farm may not, or not to the same extent, work for you. Treat their opinions and performance figures with respect, but still with caution.

   But in the areas of ease of trading, reliability of supply, product quality, adherence to biosecurity constraints and general consumer satisfaction, then the opinions of other farmer contacts are certainly valuable. Pig discussion groups and benchmarking meetings are an excellent way of picking up both good and bad vibes about commercial firms, but remember that human nature tends to be rather more forthright and opinions become embellished (either way) when people get together in a group situation rather than when given at a one-on-one walk round the piggeries.

   This go-see policy with other users is absolutely essential when researching and buying equipment, however, as the farm-similarity effects are much closer than with feeds.

- ### *Trying it out*

   Of course the acid test is the performance of the food itself. The manufacturer's own results should be treated with caution too, but not with suspicion, because the test farm conditions, including independent research farms, could be quite different in many ways to yours. Feed firms don't doctor the results, but they do publish the better ones and are tempted to leave the rest in the drawer. I deal with this in the section on On-Farm Trials (p 99).

By all means do on-going trials with a competitor's food, but get each one statistically designed and the results realistically analysed. You'll be surprised what you *cannot* say about the results, even if at first examination they turn out to be encouraging to your (and my) innumerate eyes.

Blaming the feed!

'Blame the feed' and 'blame the vet' are the two longest running saws (perhaps that should be sores?) in pig industries worldwide! The most common complaint is "The pigs don't seem to eat enough these days of that last batch you sent."

If you look at the table below, you will see that only 6 of the 18 main factors affecting appetite might be laid directly at the manufacturer's door, and these the compounder can quickly assess for you.

FACTORS AFFECTING DAILY VOLUNTARY FEED INTAKE

Pig factors	*Environmental factors*
• Bodyweight;	• Temperature – a very important factor;
• Genotype;	• Ventilation – another very important factor;
• Sex.	• Humidity;
Feed factors	• Poor lighting;
• Physical form – pellets or mash.	• Dust and noxious gas levels.
• Pellet quality. Intakes up to 10% greater if pellets are fed	
• Energy density – pig has ability to adjust DE intake to a degree;	*Pig pen design and management factors*
• Fibre content;	• Water supply
• Lysine level and balance of other amino acids;	• Pen group size and stocking rate;
	• Pen design and shape;
• Palatability;	• Feeder design and placement;
• Freshness / mycotoxins.	

The effect on pig behaviour patterns can have major influence on individual pig feed intakes.

Many of my 25 years in the pig feed manufacturing trade working for two very different companies, involved following up 91 cases of 'blame the feed'. I kept records. In 72% of the cases a very likely reason was found for the poor performance, and 28% were largely 'unsolved' – no clear reason emerged for the problem. Of the 72% where the problem was identified and sorted out, only 1 in 9 of the cases were directly or probably the manufacturer's fault. The fact that I was a pig farm management specialist probably made me the best choice to get to the bottom of a 'blame the feed' problem, rather than a nutritionist being sent to the farm. (Could be that their time was more valuable anyway!)

Quite a few of these cases were due to just feeding he wrong diet as the following table shows.

EXAMPLE: CLIENT FED A LOWER NUTRIENT DENSITY GROWER/FINISHER TO THE SAME GENETICALLY-IMPROVED PIGS ON COST GROUNDS. DIET COST £13 TONNE EXTRA (+8%). Conversion rates: £1= $1.82, €1.47

Nutrient density	FCR	Production per tonne of food (kg)		Value/tonne (£)		Difference/tonne (£)	
		Live weight	Dead weight	Live weight	Dead weight	Live weight	Dead weight
High	2.4	417 kg	321 kg	£313	£334	+£24	+£37
Medium	2.6	385 kg	294 kg	£284	£297	−8.3%	−12.5%

Despite costing £13 more per tonne, the better quality diet was worth, in cost per tonne of feed equivalent, this much more…
… **Liveweight terms** £24 – £13 = +£11 ($20, €16)/tonne
… **Deadweight terms** £37 – £13 = +£24 ($44, €35)/tonne

Not only did performance suffer because the wrong food was fed, but any cost savings/tonne were easily overtaken by an improved income, just by suiting the feed to the genotype.

Revealing this type of (inadvertent) mistake is a common part of my work as a consultant today.

To summarise…

• What the textbooks don't necessarily tell you about buying feed is how you go about securing a supplier whom you can trust.

• To do so you need to ask a lot of questions of their nutritionist, and so gain that confidence in them – as well as letting them know you have your eye on them.

• And supply him/her with ample data about your farm, your pigs, your market and your own-grown or bought-in cereals.

• Buying food well is a two-way process these days and has gone far beyond accepting what any feed leaflet claims or what the feed rep is told to say, which was what purchasers tended to rely on 15 years ago and which my advice of the time was given on how to check up on it.

• Negotiate the best price/deal, of course, but don't get hung up on price. Avoid this by always relating price per tonne to the MTF figure multiplied by the price per kg obtained for the saleable meat. This often gives some surprising beneficial results on a seemingly expensive food – and sometimes for an apparently very cheap one too. Read the New Terminology section on MTF and FCR for more details.

BUYING BREEDING STOCK

In contrast to the preceding piece, this column on how to buy breeding stock has stood the test of time well, and in my opinion needs no alteration apart from an update on the list of diseases prevalent in Europe in 1989. Today I'd say that the mention of cooperation with your own veterinarian would automatically cover that aspect anyway, as not all diseases are common to one breeding locality necessarily.

In the ensuing 15 years since this item was written, two major pig breeding companies wrote to ask permission to use the notes as in-house training material. This was gladly given; I was flattered.

What are the questions you should ask yourself, and then salesmen, when buying breeding stock?

Check your prospective multiplier: In most cases you will be receiving your stock from a pure or cross-bred multiplier, not the nucleus or nucleus multiplier the seedstock firms like to talk about. Check him out – ask for names of other breeders who have bought his stock and telephone them. More important, search out others who have *not* been recommended and ask their opinion. Once the breeding company's multiplier has been agreed as your supplier *go and see him* and learn how he prepares 'your' pigs for entry into your conditions. You may not get to see the piggeries or the pigs due to biosecurity constraints, but he should certainly have a video to show you. He could send you a CD Rom to save you the bother of travelling. Say "Thank you, but no; I want to talk to you personally".

Ask for the Vendor's Conditions-of Sale Document: These, like the multipliers, vary enormously. Read them carefully. If you don't like parts of them, say so and threaten to go elsewhere unless the condition is modified in writing, or explained satisfactorily. Check closely what they say (or don't say) about animals which do not breed satisfactorily.

***Check the General Difference between the Lines*:**   Noticeable differences are appearing in the genetic strains coming from various seedstock houses i.e.. conformation, appetite, lean-gain efficiency thus type of finishing food advised, docility and mothering qualities, leg strengths and hyperprolificacy. There are also quite major differences within the breeding companies own line structure, so check that you are getting what your market or system of production needs, not what they think you want, or maybe is convenient for them to sell you. For the progeny of outdoor sows, concentrate particularly on **proof** of fast, lean growth, as despite what is said, this could still be a weak area compared to white indoor breeds.

Breeds (genotypes) do vary considerably, even today. A good appetite is vital in ad-lib growers, and in lactating sows, especially those with large litters where weaning is later, and in hot conditions. Table 1 shows the type of variation encountered and Table 2 follows this up when I compared the strain with an excellent intake reputation with one less endowed.

Table 1.   APPETITE AND LEAN GAIN PER TONNE OF FEED OF 4 MAJOR EUROPEAN AND AMERICAN BREEDS (PIGS 25-100 kg)

	Appetite (kg/day)	*Lean produced/tonne feed**
Breed A	2.84	301
B	2.78	268
C	2.79	274
D	2.51	253

* *Same standard feed fed, but adequate diets fed* ad-lib *throughout.*
<u>*Source*</u>: *RHM (unpublished)*

Table 2. PIGS 27 – 103 kg

	Appetite (kg/day)	*Lean/tonne feed**
Breed A	2.82	309
Breed D	2.49	317

* *Breed D's feed was 8% more expensive but 11.7% less was eaten giving virtually the same yield of lean.*
<u>*Source*</u>: *Gadd (Clients' Records)*

Table 2 shows that ***where fed the right diet*** there was little difference in yield of valuable lean meat, and when the savings from the cheaper diet of Breed A was included in the profit figure, there was even less difference. So breed-specific diets are important, called 'Feed for the Breed'.

In my experience it is important when choosing seedstock to balance appetite potential with dietary cost and some breeding firms (not all) are now getting much better at providing this information.

So in addition to the three main areas listed below you should ask about *proof of a good appetite* together with the nutrient specifications of the diet to match whatever it may be.

Get to know the vendor's salesman/pig specialist: Once you have established a relationship based on trust (this takes time) he or she will be a key factor in ensuring you get the right stock for your system of production **on time, checked personally by him/her and old/well-grown enough**!

Here are some routine questions you should be asking of all seedstock salesmen

Record carefully how they reply and what the answers are – it helps you decide when the choice is wide, as it is these days.

GENETICS

1. a) What proof have you that the performance of your herd is normally better than average.

   b) What schemes do you test under?
2. What selection intensity do you use on the farm?
3. How do you ensure that the variation you are measuring is due to genetic ability and not to variations in management?
4. How quickly does the farm replace its females?
5. How many years have you been making consistent genetic improvement?
6. How large is the genetic pool from which you select?
7. What proof can you give that the prospective breeding stock is free from genetic defects?

HEALTH

1. What proof can you give that your stock is healthy?
2. Can you provide independent evidence, such as from a veterinary surgeon's reports? Are you free from such diseases as: –

Enzootic and Haemophilus pneumonia, Atrophic Rhinitis, PRRS, Swine Flu, Aujeszky's, Transmissible Gastro-Enteritis, Swine Dysentery, Swine Fever, SVD, Leptospirosis, Meningitis, Worms and Ectoparasites?

3.   Are any drug treatments given as a routine which would mask the symptoms of any undesirable condition, for example, Atrophic Rhinitis, Swine Dysentery, Meningitis, *etc?*

4.   Will you give permission for your veterinary surgeon to give me a full history of your herd? And for him to compare notes with my own vet?

Note: *This list refers to main European problem diseases 15 years ago – it may be different outside Europe and your veterinarian will update you on the local position anyway.*

PROFITABILITY:

1.   Can I purchase your boars/semen on an EBV system? Will you help me to chose the most cost-effective of your boars for my particular circumstances – whatever their price?

    Note: *Today, the same applies to AI semen.*

2.   What proof can you give that the prospective breeding stock are prolific, sturdy and will thrive under commercial farm conditions?

3.   Will your slaughter pig progeny sell at an above-average price? Proof please.

4.   Are your pigs likely to show a better than average profit?

5.   Are your pigs good value for money? How would you compare them to your typical competition?

Generally, seedstock companies do a conscientious and good job and should be able to answer positively to all three sets of questions. But the fact that you have asked them puts them on their mettle and signals that you are a priority customer. And an astute one!

BUYING SKILLS

Another column which really doesn't need alteration is this one on buying skills, written towards the end of 1989. Some pig producers will question my statement made at the time that profit depends two thirds on maximising output and one third on minimising costs, and that today the importance of reducing costs in most pig industries predominates.

"Well," as the sage said, "It all depends!" It depends in this context on how good is your production performance in the first place, doesn't it?

To explore this subject further, and how the two might be brought into comparison on any farm and acted upon, the New Terminology section covers this aspect.

Agricultural Colleges please note:

Fifteen years ago I criticised the farm student training centres (as distinct from degree, masters and veterinary courses) for not training stockpeople and especially future farmers in *basic* statistical analysis so that they could better understand (and assess) what advertising and the sales patter was claiming. Having made half a dozen phone calls recently I suspect it may not be all that much better today. 'We touch on it' was the best response I could get, and several said that the syllabus was so crowded that there was little or no room for it to be covered to the extent needed. Oh dear!

This is a numerate age. Pig farmers still tend to be numerate illiterates. As I once was – and comparatively speaking with the scientist, I still am, I suppose. But knowing I was weak in this vital discipline – because that is what it is – stimulated me into looking afresh into this whole area of the way pig farmers need to approach numbers, measurement and nomenclature in what they do. What I've done about it I describe on pages 39-46 (*New Terminology*).

I suspect that our agricultural colleges still need to pay much more attention to training would-be farmers and managers in how to look at figures, theirs and those of others. Meanwhile, there is a fast-track route to getting it right which I outline on page 103 (*Using a Statistician*).

There isn't a single textbook on pigs I've read which even starts to approach 'how to buy things' – pigs, food, equipment, drugs. Yet every pig producer has to buy over 85% of his input costs. And while profit depends two-thirds on maximising output and one-third on minimizing costs, why is it the bad buyers who go under first? This is because a failure to reduce input costs has such an immediate influence on cash-flow, and it is usually an inability to support a viable cash-flow which finishes a pig business, not necessarily eventual profit.

Farmers are not nearly as good buyers as they think they are. Even today much of it is emotive and instinctive. Neither should come into it.

Here are some basic rules of good buying.

Make time to buy well. Some people tell me "I rarely see reps – too time wasting". This is unwise. You should put at least 10% of your time, say ½ hour/day, to talk to vendors. If you don't, how can you negotiate properly? How can you reconnoitre the market place properly? Perusing ads and visiting shows is peripheral – you must make time to seek out what approaches are on offer, and acquire enough hard facts to play one rep off against others. But always give a rep a timed period to discuss his proposal – 15 minutes is adequate.

There's strength in numbers! Commercial vendors love individuals, they are more vulnerable to persuasion and the margin is higher. Anyway small orders do justifiably cost the manufacturer more. So seek every opportunity to co-operate with other producers so that regular farm inputs –food, disinfectants, drugs – even veterinary supervision etc – can be negotiated from a position of co-operative strength. This also enables the vendor to plan his logistics so that he can reduce his costs to you, and still reach his budgeted profit.

Always look at leasing, not buying. This is often attractive especially when interest rates are high – and it can help cash flow. Mathematically the decision is easy – it either is or isn't cheaper over a period. What is more difficult (in machinery and equipment) is when and how much your part-share will be made over to you and (pig breeding stock/food) to what degrees of restriction you are subjected. In reality farmers make more fuss over this than is justifiable as much can be amicably negotiated within the surface small-print when competition in the vendors field is high, as it usually is or leasing wouldn't be offered.

Lies, damned lies – and statistics!

A much-maligned phrase. Most salesmen don't lie – they are trained to magnify the good points and skilfully minimize the disadvantages of their product or service. To get nearer to what is likely to happen on your farm you have to know what questions to ask.   These differ substantially according to whether it is food, or growth promoters, or vaccines, or gilts, or boars, or equipment. In fact my own talks on 'How To Buy Food', 'How To Buy Breeding Stock', 'How To Buy Ventilation Gear', etc, are extremely popular worldwide and there is not room, of course, to outline them in one column, so see the relevant sections in this book.

But there is a common theme running through them all, and this is: -

GOOD BUYERS: -

•   Have done their homework; they know what (awkward) questions to ask; they know the product's market; they know the competition involved.

•   Are aware of basic statistics when claims are put before them. Things like confidence limits, repeatability, significance, the effect of variables etc, are of vital importance in assessing the value of what is claimed over what you are achieving now.

        Sadly, most farmers will continue to be duped on this score until they are trained in the simple appreciation of what the presented figures really mean, and sadly, vendors will continue to offer half-truths dressed up impressively until they find that the buyer can soon spot the flaws – that it may well not happen on their farm. Colleges, universities and extension/ training services in many countries could be failing the working farmer in this area. They say to me that the farmer cannot understand simple statistics – but this is an admission of defeat/poor training ability – and wrong!   After all – *I've* had to learn the hard way!

•   Have identified their priorities – or got help in identifying them – what products on offer contribute most to overall profit. Food, warmth, space, labour, disease? Which are their/your priorities?

•   Know clearly – and always bring the seller back to this core data -what their end-product is. For example, in the case of food it is value per kg of food of the quality lean meat sold. (We are *meat* producers not *pig* producers). In the case of breeding stock it has to be split between boars/AI and sows (male and female traits are so different) so evidence of Estimated Breeding Value is essential, then knowing how much to spend and where to spend it, in both categories. In the case of growth promoters or their alternatives it is sufficient statistically valid evidence under conditions as close to yours as possible. In the case of equipment it is true capital maintenance and depreciation costs at the current interest rate plus agreed running costs (if applicable) divided by the extra output expected over the current/circumstances. In the case of housing it is proven output per metre^2 set against all costs and funding interest needed, including labour and slurry removal.

At the end of the day good buying is not a copy-book process. But it depends heavily on establishing as many hard facts as possible and percipient questioning so as to balance likelihood against quoted cost.

Cheapest is not always best. It is cost-effectiveness that matters in the end.

Footnote:

Talk in realistic terms – use analogies. David Taylor, my old employer, always used analogies to bring home a message to prospective vendors. I listened and learned from him, and since then have taken this particular leaf (of translating some otherwise obscure figures to something we can all understand) out of his book to help drive a point home.

Let me give you an example from our daily lives. We all read newspapers where the media continually mention businessmen's 'fat-cat' terminal bonuses in terms of millions of pounds/euros/dollars – whatever. *Such telephone number sums mean nothing to the buying public*, so I did an exercise one day, when I discovered by chance how many customers one international company had and what their sales were in a year. This was 17 years ago.

It was easy enough to divide the gigantic terminal bonus the CEO got (millions) when he was sacked (presumably for ineffectiveness), by their number of customers and that amounted to £2.36 ($4.30; €3.47) for each customer! Including myself! That would be nearer £6 in today's money ($10.90; €8.80). As I'd only spent £30 ($54.60; €44) with them that year, 17 years ago, in my case it added up to a loading of 7% of what I'd paid them – all into the pocket of just one man (who apparently was a failure) and not to the firm itself as profit, which would have been fair enough.

If the media (press and TV) would only relate these ridiculous bonuses to what they *really* cost you and me in pounds and pence/dollars and cents instead of talking about millions, we'd see less of them, I guess, as shareholders and customers would get *really* angry and stop it. This hasn't happened yet.

As a result of this exercise, I've never spent a single penny with the firm this past 17 years, and never will again if they are going to waste my money spent with them in this way. Their products are good – but to blazes with them!

SELLING TO THE PIG FARMER

Much of my commercial (employed) life has been engaged with training salesmen in pig technology, and I even did a bout of sales representation myself. I thought it worth doing a piece on selling to pig producers, not only to assist the salesperson in his/her difficult job, but also because I thought the pig farmer might be interested to read how that friendly salesman is being trained in how to handle him!

Poacher turned gamekeeper? Sure - why not!

I have now survived 42 years of involvement in the pig industry. Yes, survived! Anyone connected with pigs knows that survival is very appropriate in our circumstances. And anyone interested enough to read this must be a survivor, too.

Half of this long period was spent in commerce, either being directly or peripherally involved in selling and marketing products to pig producers (selling and marketing are not the same things).

There are many, many textbooks and manuals on selling technique, but there are none that I know on selling to farmers, and certainly none on persuading the pig producer to buy from salespeople. During those 40 years I have been fortunate to sit at the feet of several brilliant men and women in their own careers inside and outside of agriculture. Two of them were highly successful field salesmen; their annual sales ran into millions of pounds sterling in one case, US dollars in another. But their persuasive styles were quite different and their approach personalised to each customer, so there was no real point in copying them – only some basic techniques common to both, which were, when I reflected on them, largely commonsense anyway.

One day, not so long ago, I listened to a presentation by one of the most successful company Chief Executives in our field of animal production. Virtually single-handed at first, then with a team of dedicated employees, his firm has become one of the biggest businesses of its kind after only two decades.

No textbook has put across the theme he developed at this meeting and, although his remarks were aimed primarily at the agricultural salesman, the final target of all his advice, the pig producer, will be reading this and should find interesting what this brilliantly successful 'salesman-extraordinaire' told us. We can all learn from it, on whatever side of the gate we are.

Getting appointments

- Phone, but confirm by writing, faxing or email.

- Keep to schedule. Farmers are busy people and breaking/delaying appointments can irritate. Conversely, being recognised as a good timekeeper gains confidence.

- If you are running late – phone through.

- From time to time, bring a little gift. Not those silly commercial 'giveaways', but if it's a hot day, a can of Coke. A cold day, some doughnuts. Or "Saw this leaflet at the Show last month and thought you'd like a copy".

Get to know who is the boss

- Are you targeting the guy who signs the cheques?

- Get to know who the likely replacement is in case the boss leaves, and befriend him, too.

Be happy, be fun

Smile. People like to work with successful personalities.

A glum mood when you arrive? It is tempting to sympathise with adversity. Be careful, it can leave a negative impression, even so. Like as not, the reason for it will be a low pig price, disease, staff problems and red tape which are the main causes of po-facedness these days, so prepare in advance some positive suggestions just in case.

Meet outside his office if possible

Sure, the records are stored there, but the upfront conversation benefits from where the pigs are – unless it is feeding time! Why outside the office? Far fewer interruptions !! But these days, mobile phones are a nuisance!

Don't sell products, sell solutions

Think how every product you are asked to sell can be a solution.

What are your pig processor's problems? As a farmer how can your pig production business help solve/lessen them?

Keep good notes

Allow time to pull into a lay-by and record, not just the major items, but what are his preferences/dislikes/family details.

Comment: **I'd like to mention my own experience in this respect. I've always had a terrible memory, so as a young pig adviser working for a large feed manufacturer, I disciplined myself always to note down minor details so I could surprise and impress the client should I ever call again – sometimes (in advisory work) as long as a year or two later. This soon turned into a daily farm diary, later a general 'Omnium Gatherum' journal (now some 37 years old and 3½ million words, 36,000 photos in over 100 volumes).**

As a result I have a reputation among pig farmers for having a photographic memory! Not so. Not so at all. Never have had! All I did was to look up the Index of when I last visited and remind myself of the trivial details like the youngest daughter's name and age, whether the guy did get £80 ($145; €118) for that cull sow, etc. - and brought them into the conversation when the opportunity arose.

Agree with the hard-to-get customer

He may be right anyway! Agreement is a definite 'softener'.

Grade your accounts

A   Key account. Your easiest sale is to an existing customer

B   Immediate potential. Frequency of attention/visits pays dividends.

C   Hard to get. Persistence pays with four out of five of these. You may have to be patient so don't give up.

D   Prospects yet to call upon.

Divide these categories into how much you have got to sell to each one to achieve your annual target/budget, and spread your time/effort accordingly on a weekly planning basis.

Make a plan – stick to it

Based on the above general strategy, make a monthly route map of calls.

Always follow up immediately

With a letter, fax, email or phone call. Mention only three points, no more. Keep your name/firm's name in front of him between calls.

Don't assume anything

Things *do* go wrong. Orders, help promised by other colleagues, slow responses, the post, messages not passed on. Check, check, check. Did my letter/quote arrive? Call to ensure all is well. Show you care: make time for this.

Never run your competitors down

In fact, never mention competitors; this is totally negative. If the farmer does, show courteous interest, no more. A lower price is often involved, so have your response ready prepared.

Prospects

You will lose 10-15% of customers each year at least. So you need to replace them.

Farmers have to sell their product, too. While the above sound advice is aimed at the agricultural salesman, in a one-on-one situation, there are lessons in there for the pig producer, nevertheless. He has to sell his produce, lean meat (not pigs) to the processor, or direct to the public. Making it easy for them to buy could help get him a premium price. And paradoxically, he also needs to 'sell' *his* **problems in producing pigmeat, to the very people who sell things to** *him* **– so as to secure a better price, product or service.**

So what could be your selling points to your immediate customer, the processor? Consider the following…

1. He wants a *consistent supply* of *contracted pigs* (i.e. you will supply him with the number of pigs which you have both budgeted for *on time*) *within agreed weight bands*.
    All these reduce his costs and give him no trouble.
2. You can supply a pig with a *high yield* of *good quality lean meat*. Make sure he knows the effort and extra money you put into *customer-friendly genetics* – in this case the final customer is the retail outlet.
3. Again *the right food to achieve this* is actually more expensive per tonne. Translated into per kg of liveweight/deadweight, this can be as much as 1.5 pence (2.25 cents) per kg liveweight.   Work it out in your case and make sure he knows you are *investing this on your joint behalf*.
4. *Cleanliness of production* is high in their priorities. This is not just providing washed pigs to the abattoir, but producing them hygienically. If you do this, invite them to visit and see for themselves. This involves *salmonella control* and *lung scores* as well.   What are you doing to achieve this?
5. *Show an interest in his problems* by visiting occasionally, asking to look at your carcases. Keep your face before him.
6. If selling direct to the public, concentrate on the 4 Cs… *Clean Food*, *Clean Conditions*, *Comfortable Animals* and that y*ou Care* about what you do.

Farmers should remember the Golden Rule of Selling.

… Make it easy for people to buy my product.

QUESTIONS TO ASK THE FEED SUPPLEMENT SALESMAN

... Now let's move to the other side of the gate, the farmer's side.

Even the busiest pig farmer or his manager should find time to listen to the salespeople who call. When I was farming I put by 15 minutes per sales visit (totalling about 45 minutes per day) to talk to them. Fifteen minutes per call? Any well-trained salesman should be able to get his message across in 10 minutes, and I always told them so. This left 5 minutes for 'gossip' about local conditions, how other customers were doing and the pig (or feed/pharmaceutical/ equipment) trades in general.

It was these 5 minutes which I found the most interesting and useful to me in my job. I rarely refused to see a salesman, even when busy. They might have to wait a bit or come back later, but I always gave them listening time. It kept me in touch with the commercial world.

Anyway.... As AGP replacements is such a topical subject at the time of writing, let's look at this subject under the 'Questions to Ask' heading.

Antibiotic growth promoter replacements

Today, so many salesmen seem to be pushing their own brand of AGP replacements. I count at least 20 on the market and I expect there are many more. All very persuasive. All very confusing. But which to choose?

I was therefore intrigued by a paper my old friend of 30 years, Dr Gordon Rosen, gave last year when he listed seven questions to ask people selling AGP replacements.

Dr Rosen is a highly qualified and respected scientist and worldwide consultant to the feed trade. His approach to the statistical accuracy and validity of research into pig and poultry nutrition is exemplary. He has carried out several 'holo-analyses'[1] on what such results really mean to us at the sharp end of on-farm advisory work, and is a stickler for descriptive accuracy on what you should and what you should not say about the efficiency of properly-controlled trials.

---

[1] A holoanalysis is a comprehensive analytical survey of all available test results and variable

His seven questions complement, and improve upon, my own earlier suggestions on '*How to Buy Feed*' and '*How to Buy Breeding Stock*' from salespeople in those areas who call.

Using the Seven Questions Test (7QT)

My conclusion is that the farmer should assess the likely value of each product being promoted on how far the manufacturer/distributor can go towards satisfying these quite stringent requirements of Dr Rosen. Obviously in some cases – for example as in the case of a potentially promising product which nevertheless is relatively new on the scene – the volume of data may not (yet) be there.

I myself have tested out the 7QT – nominally – on several products, and, in a comparative sense they did sort out 'the men from the boys'. Even if the answers were not as complete as Gordon would hope for, they did help shorten the list of possible products. As with any judgemental decision on a choice of options, one needs to assess things objectively and Dr Rosen's 7QT, as he calls it, certainly does that.

The Seven Questions

1. **How many properly controlled feeding tests do you have on your product?**

    Dr Rosen feels 20 such tests are needed for a first appraisal, but more (50+) may well be needed to account for key variables.

2. **How many of these tests have no negative controls?**

    A negative control determines whether either product on test is effective in a trial by comparison. Having just a positive control must limit the value of the trial because it does not show whether either of two products compared is effective. All trials need a negative control.

3. **Have you a list of references to support the first 2 questions if required?**

    Comment: This shows where and when trials were done whether in-house or by an independent service. In-house trials can and have been selected to publish only the favourable results. I always ask if *all* the trial work done is available, negative as well as favourable. I suspect 'all-positive' evidence. Even with a good and valid product, one gets failures. So...

4. **Response frequency. How many times out of 10 does the product improve performance?**

Dr Rosen notes that over the years the effective and acceptable pronutrient antibiotics have 70% - 75% response improvements and any replacement reaching this standard could be considered acceptable. 100% would be suspicious!

5. **Response co-efficients of variation (CV). What are these?**

These reveal the amount of variation in responses over a range of use conditions, so the smaller the better. Variations of up to ±50% should be acceptable.

6. **What dosage of the product will maximise return on my investment? And why?**

Comment: Applying econometric indices to performance-related research is vital, as I myself have been explaining to clients for 20 years, even having to invent 'New Terminology' (REO, MTF, AIV, etc) to make this vital application of worthwhile research intelligible to these where capital is limited and choice is great. *Always* ask a supplier what is the return on investment at half or double the recommended level – these are extra data useful in coming to a conclusion.

7. **Can you supply me with models to predict responses to your product under my farm conditions?**

Modelling for liveweight gain, feed conversion and carcase responses is still an under-used technique by researchers, suppliers and by farmers, and while models exist for antibiotics and enzymes at the time of writing, they are much less common as yet for other replacements, such as organic and inorganic acids, micro-organisms, botanicals, etc.

Comment: The modelling does not only enable comparisons to be made between the performance-promoting products themselves, but also can be put alongside other options (housing improvements, equipment replacement, dietary alterations, including extra nutrients, etc) to compare likely added value potential.

Are these Seven Questions of practical use to the pig producer?

Certainly, if interpreted properly. This is best done by recording the salesperson's ability to answer them. This initial screening process will quickly reveal those that are highly questionable and rejectable. The remainder should be referred to a statistician or trials officer at a local college or animal research station – for a fee, of course. These fees are not, in my experience onerous, and in examples I have written about elsewhere (p 103), have been worthwhile.

After all, for the finishers produced by each 100 sows in a year the producer is likely to be investing at least €1300 (£884, $1,609) annually on his chosen replacement pro-nutrient feed additive. Getting a consultant, trials officer or statistician to tell you what you can or cannot believe about the manufacturer's or salesperson's claims in my experience has only cost a fraction of that. To choose largely 'blind' (i.e. on plausible but nevertheless insufficient evidence) is not a good use of scarce capital.

So we need to question these salespeople harder.

Equipment and housing

My remarks so far cover nutrient additives and whole diets. But what about say, ventilation equipment, mainly fans? In this area I find pig producers very trusting, and at times naïve. Fans and ventilation gear tend to be bought either locked in to a housing package if a new build is envisaged, or 'because it looked good at the Expo' or 'the specification figures seemed impressive' or 'I know the firm/trust the vendors', if replacement/refurb. is required.

As in pig nutrition, in the field of ventilation equipment we are dealing with a technically-precise subject. Nutrition is biology and mathematics, ventilation is physics and mathematics. The biology and physics may vary a bit as research hones and refines current knowledge, but the mathematics are ruler-straight! They don't brook argument.

Because of this, choosing any old fan is not on. *It must come carrying the vital mathematics specific to your circumstances, both in capacity, placement in the building and design.*

Thus farmers need to request the manufacturer's salesperson to:-

• Carefully measure and assess the structure which the ventilation equipment is to service.[1]

• Have calculated the dynamics of the maximal and minimal target pig throughput, including expected ranges of climatic change, wind and air speed, temperature degree days, RH bands, etc.[1]

• Submit plans and section drawings of the proposed ventilation system, including air distribution routes which will meet all these agreed variables, together with the fan specifications needed to do so.

• At least two, preferably three suppliers should be asked to do this. My experience reveals that *the results will often be different* and an independent ventilation engineer recruited to pass judgement is wise, if not essential.

[1] All these procedures are laid out in my Solutions to Pig Problems book pps 489-497.

SPECIFICITY AND PRIORITIZATION

Few textbooks on pig production touch on these two increasingly important subjects.

Specificity

Long gone are the days when a consultant could give even roughly the same advice on most of the farms he visits. Econometrically, each farm is quite different and the advice given is therefore varied. This is one reason why I tried to avoid writing a textbook until recently! Textbooks have to deal largely in generalities, and it is the *application* of all this wisdom in the right *place* and *time* and *order* which is the essence of good farm advice.

Table 1 gives a real-life example of specificity. Both farms are neighbours and even use the same genetics and food. In analysing their performance, as I do once a year, it is not difficult to calculate what effect shortfalls in various production sector targets may have on profit. Notice how, of the sample of six important sectors (out of about 22) 50% are radically different between the two units and merit different advice for the two farmers.

Table 1. EXAMPLE OF COST/PROFIT ANALYSES ON TWO SIMILAR AND NEIGHBOURING FARMS, IN THIS CASE VISITED ON THE SAME DAY. BOTH NEEDED QUITE DIFFERENT ADVICE TO IMPROVE PERFORMANCE AND PROFIT. (NOTICE IN TABLE 2, BELOW, HOW DIFFERENT THE COST OF RECTIFICATION WAS)

Areas which records and the farm tour revealed as sub-standard	*Effect on net margin %*	
	Farm A *	*Farm B ***
Lower early life mortality (12% down to 10%)	+17	+ 5
Farrowing rate (5% better)	+16	+18
Born alives (10.75 up to 11)	+18	+ 6
Less post weaning check (12 days down to 9 days)	+ 3	+19
Ventilation/respiratory disease (25% incidence reduction)	+22	n/a
Better meat yield (+0.5%)	+ 4	+13

* Farrow to finish, 250 sows   ** Farrow to finish, 400 sows.

Prioritization

Any on-farm consultant soon learns that the cash available to help rectify target shortfalls is limited. There is never enough spare money to do everything that the analysis suggests. So which to do first? Again, with experience it is possible to put a price to these options. Some are a lot more expensive – for example like rectifying ventilation or altering buildings to All-In/All-Out (AIAO). Trouble is, the less experienced and often enthusiastic adviser tends to go for the improvements which will secure the biggest profit improvement and doesn't include payback and cost into the calculation.

Prioritization looks at the projected *cost* of the rectification, the *time* this cost takes to payback out of forecasted improved profit and the *amount* of profit likely to accrue overall. Table 2 gives a prioritization exercise done on Farm A's options.

Table 2. AN EXAMPLE OF A REAL-LIFE PRIORITIZATION EXERCISE

Shortfalls against target as revealed in records		Estimated relative cost of getting each up to target (Lowest shortfall = 100)	Estimated time needed (years)	Predicted effect on current Net Margin
Born alives	– 2.25%	100	0.75	18% better
Early life mortality	– 5%	250	1.2	16% better
Farrowing rate	– 20%	20*	0.5	17% better
Ventilation adequacy	– 25%	2300*	1.4	22% better

* i.e. The cost of improving ventilation is 115 times more than the cost of measures to reduce mortality to weaning by one-fifth. Data from clients' records

Sure, ventilation improvement was a significant option both in performance and economic terms, but as everyone knows, is usually very expensive. Table 2 shows that the farmer (and his adviser) must relate the cost/benefit to other options which may be better value for money/easier and cheaper to implement *first*.

We can see that attending to *farrowing rate* is the first to tackle, because it is not expensive and thus the payback is short. After 6 months, and for every six months thereafter, we should enjoy a 17% increase in net margin. Measures to improve *born alives* are also a good bargain. It is still relatively inexpensive and after 9 months we might be able to add another 18% to the net margin. After 14 months, *early life mortality* improvement could net us another 16%. And so on.

By two years, say, on the examples given, we could have secured quite an improvement in cash balance, which could go towards paying for the very expensive ventilation alterations needed.

What happened in real life ?

In reality, on this farm the ventilation replacement project was held off for 2½ years (and the cash needed for it increased by 13% meantime!)

During this period, respiratory disease was held at bay by vaccination. Even so, by the time the project was started, over half of the capital sum needed was already there from the other three projects receiving priority over it.

However, had control of respiratory disease been a real problem, and discussion with the veterinarian and the agricultural engineer implicated the outdated/worn-out ventilation as the prime cause, then priority takes over from specificity and the capital investment allocated.

Specificity *and* prioritization, you see. They go hand in hand.

Allow ample time

A piece of advice. If you do have a major capital project looming – allow at least 3 years to commence prioritization exercises like the above. Some time in those 36 months will be the right time to move – maybe early, maybe later.

In my opinion, many pig farmers are too close to their own businesses to prioritise objectively; this is where a modern-day consultant can help in doing the legwork so that the options are clearly presented and the ideas on which they are based are laid out. This is the right way to use a farm adviser.

The priorities today

It is surprising how pig industries across the world – and I've worked in 21 different countries to date – are similar. Similar too, are their priorities. There is a difference, of course, in the very hot countries where keeping pigs cool is a priority most of the time compared to Canada, Hokkaido Island (N. Japan), Korea, Scandinavia and Central Europe where (mostly) heat conservation lies uppermost in their minds.

Using this experience, much of it fairly recent, I list what I consider to be the areas in which modern pig producers should concentrate. By not attending to some of these priority areas (on grounds of cost, difficulty, labour shortage etc – but not of capital; there is plenty of money available, at a price) pig farmers are making life more difficult for themselves.

I hope this brief list makes you think a bit. And I encourage you to stand back and consider where you are in each category.

Pig Price Most of us cannot do much about this. But we can think about who our customer is and are we making it easy for him to buy what we have to sell – quality pigmeat?

Disease At the time of writing some viruses seem to be gaining the upper hand. Why? This brings up a whole host of 'sub-priorities' which taken and acted upon as a group will had have made disease much less of a worry to certain successful clients of mine – past and present. What are these?

Learning about immunity and how we can get the balance right between pathogen challenge and natural protection.

Updating yourself on the new virucides and how to use them.

Using the veterinarian properly as a management assistant who monitors things, not just as a 'fire brigade'. Yes, and paying for this type of assistance.

Paying more attention to how disease spreads, especially vehicles, birds and vermin.

Remodelling the farm's strategy to break pathogen build-up. AIAO; Partial Depop; Batch production; less frenetic early weaning, etc.

Tail Chasing Most farms don't get the priority jobs done properly (breeding, farrowing, serving, cleaning, monitoring) because of urgent immediate repairs/ renewals. Delegate to specialists.

Recording Getting worse, not better! Either too detailed, too cumbersome therefore not acted on sufficiently/quickly enough.
Conversely not done at all ("No time"). There's a happy medium; *only keep essential (graphical) weekly records, make inputting easy, then act quickly on what they reveal.*

Labour Study the successful (industrial and high street) businesses and copy their philosophies and the way they handle people.

VALUE FOR MONEY

There are several good textbooks on the economics of pig production, but in my opinion none of them deal well with the complex subject of spending money wisely. In a nutshell – spending the right amount on the right areas at the right time.

As a sharp-end farm adviser I've always been intrigued by the way pig farmers invest money in their businesses. This led me, 30 years ago, possibly to be the first to coin the term 'econometrics' (or if I was not the first I must have been a very early user of the word). It correctly means 'the things which describe cost-effectiveness' and more recently has come to mean 'the measurement of cost-effectiveness'.

In my work I have had to persuade producers, sometimes to spend more, sometimes to spend less, but always how *best* to spend what liquid capital they had available, borrowable and feasible. Sometimes too, it was unexpected advice, and occasionally unpalatable. Rather than being shown the door, I had to persuade him/her (and occasionally, Heaven forbid, a committee!) before it got to an impasse between us that there might be another viable – or just as important, *interesting* – option with which to hold their attention until I could convince the client that the alternative solution might be worth consideration.

Easily the best way of doing this was to use the econometric approach, but I needed a whole set of new terms to support it.

Why econometrics?

- Money can be scarce, and even if available, is not cheap.

- There are a huge number of options on how best to spend the money. To take just one example from pig nutrition, up to 2004, in my travels I have discovered at least 92 feed additives available on the world market, all claiming to improve pig performance in maybe 25 different ways!

- Econometrics narrows the scope in which an option in one area looks the best one to choose to meet profit targets or reduce impending loss. This approach can be used, not only within one sector, but to compare different possibilities between sectors – housing, feed, breeding, AI, meeting regulation requirements, safety and pollution, etc.

Because econometrics involves costs and income, I found the farm data from which I needed to get an idea of value for money to be inadequate because they were all based on physical performance, such as food conversion and daily liveweight gain, and some of them were highly unreliable anyway, like food conversion because it is so difficult to measure accurately on the working farm. How inaccurately is given in the Table on page 37 and on page 41.

I couldn't use the farm information presented to me sufficiently, or confidently enough, to convince my clients how *best* to spend their liquid or fixed capital.

New terms needed

To replace the disadvantage of having to use only physical performance terms when coming to a best-value conclusion – the method which most people still use today, in fact – I needed to invent new terms which would *still encompass performance but also introduce built-in costs and income as much as possible*. This is because you can have very good performance and yet make little money, and some quite modest performance results can retain a lot of the original outlay as profit. This second option (my new term for it is SLC – producing the **S**ame at **L**ess **C**ost) is less favoured in modern pig production, still harnessed as it is to the runaway train of ever producing more. But SLC when calculated shrewdly is a much better way of using money. As I said earlier, in spending the right amount at the right time – in SLC's case, in the right way. And sometimes in not spending it at all. Yes, that can happen, for example, being better to spend it somewhere else.

So these new terms of mine also favour *prioritization*, a vital economic item of support.

Done to make my job easier!

These new terms, based on profit, not necessarily centred on performance, were invented (actually they grew on me gradually as different problems appeared over time) to help me convince producers how to get round their poor-profit problems, and that my suggestions might be better than their instinctive solutions!

Or even, occasionally (whisper it quietly), those of some other advisers! This sounds impossibly pompous, and I suppose it is, but I have no way of saying otherwise. Table 1 summarises just some of the new terms and how they fit in to a modern lexicon based on **profit**, not so much **performance**. If you need to study the thinking behind the curtailed advice in Table 1, I'm afraid you'll need to get a copy of my 590 page book[1] where in 27 pages of a 98 page Business Section the arguments for the new terms are put forward. A very brief summary of it is given overleaf.

Supporting clients in trouble

During the dreadful pig crisis which beset my countrymen for 3 gruelling years across the turn of the century, I spent many hours with my clients in front of their bank managers. They needed help – or many of them did – in extending loans/ borrowing a little more in order to ease the cashflow problems which threatened to engulf their pig businesses

Often I never charged for my time as I felt that they were already in enough money trouble, and, after all, they had paid me over several previous years when times were better, so this one was 'on me' so to speak. But that's by the way.

Anyway I learned a lot about the way bank managers think – employed by different banking companies in a variety of rural towns – when faced with a customer in trouble. I have (now) the greatest admiration for (most of) them. Invariably they were supportive and kept us afloat – most of us anyway.

My job was to show that the farm was reasonably efficient and that we were doing all we could to cut costs, and what the effects in this direction were – in progress now, not just promises. And that when the clouds rolled away, the pig business would be leaner and stronger as a result of our emergency actions.

Reassuring lenders

Value for money, in other words. The bank managers (or the banks' farming finance advisers) were looking for evidence that the loan extensions could be justified to Head Office in terms of better efficiency as well as belt-tightening, and that the negative cash flow was due to outside factors not under our direct control.

---

[1] 'Pig Production Problems, A Guide to Their Solutions', Nottingham University Press, June 2003. For details visit: www.nup.com

Even in this worrying, negative, scenario, we were competing with other farmers – not just pig farmers – who hadn't put up such a positive case. I knew a few, and the reason was not laziness or unconcern on their part, but the gut-churning worry which stopped them thinking calmly and positively at their interview – and more importantly *before* it.

We succeeded (mostly) when they didn't because we approached the subject in a positive, constructive manner. I like to think that even in this crisis, we were seen to give (better) value for the money advanced by the bank than some others.

An experience – and a lesson never forgotten. Which only goes to show that you are never too old to learn.

I learn something new about pigs every day,
... something really important once a month,
... something of revolutionary impact once a year.

In conclusion....

There are many ways of spending – or wasting – a dollar. I hope this piece has whetted your appetite to explore other possible routes to using that dollar more effectively.

I'll go further. I'd like to see every pig textbook have a chapter on Value-For-Money on the subjects they cover. And maybe, all research reports too, when revealing a performance improvement/reduction. In the form of what does the benefit, or otherwise, mean in econometric terms based on agreed and current monetary assumptions? If you don't agree with the author's financial assumptions, you can always put in your own figures.

Too much to ask? Up to now, it seems so. A bit worrying that it hasn't happened yet, I guess? Don't you think so too?

TABLE 1. A THUMBNAIL SKETCH OF SOME NEW PROFIT, RATHER THAN PERFORMANCE, TERMS

The new terms move you away from...	Why?	Into a better term...	Why?
In feed terms – cost per tonne (ton) – Food Conversion Ratio	Unit price can be: Misleading / Unreliable on the busy farm as accurate measurement is difficult.	MTF (saleable Meat per Tonne of Feed	Sets your main cost (food) against your income (saleable meat) from sales dockets. Also encompasses FCR anyway, and more reliably.
In feed additive terms – cost per bag	Does not encompass usage rate	REO (Return on Extra Outlay)	Much better than unit cost, or cost/tonne/sq metre/pig, etc. Puts product performance and overall cost into perspective
In sanitation terms – cost per bag or drum	Does not encompass dilution rate or surface cover	REO (as above)	
In housing terms – cost per m^2 (or in the USA ft^2)	Too restrictive, no long-term guideline. Base cost tends to put people off.	ILR and PLR (Income to Life Ratio; Profit to Life Ratio) AIV (Annual Investment Value)	These look at spending over the whole term of use and are related to improved performance. Looks at spending in the same way a lender does. Good if you need to borrow money from him.
In growth rate terms – cost/kg gain. (Can still be used for overheads, etc)	Only reveals costs, not income	MTF is better than cost/kg gain	It is more comprehensive – tells you more.
In breeding terms – cost per boar, gilt or AI dose	Short-term, can be misleading	REO (see above) and EBV (Estimated Breeding Value)*	Both these terms are very useful in comparing breeding stock on offer.

A THUMBNAIL SKETCH OF SOME NEW PROFIT, RATHER THAN PERFORMANCE, TERMS (CONTD)

The new terms move you away from…	*Why?*	*Into a better term…*	*Why?*
In monetary terms – borrowing costs (e.g. depreciation/Return on Capital)	Tends to be off-putting to both lender and borrower.	AIV (see above) ROI (**R**eturn **O**n **I**nvestment needed – not just return on capital) SLC and MSC (producing the **S**ame at **L**ess **C**ost; producing **M**ore at the **S**ame **C**ost)	All are much more intelligent ways of looking at the need for borrowing money. Much easier to convince lenders to increase or extend loans. Gives the impression you have thought things through.

Similar alternative terms can be used for veterinary, labour, mortality costs to replace the common and potentially misleading performance terms used today.

EBV not invented my myself. The rest, and more besides, are.

REO (Return on Extra Outlay) is very useful when comparing alternative options, also expressed as R.E.O. The highest REO wins in financial terms, anyway.

A.I.V. (Annual Investment Value) looks at investment expenditure in the same way a lender does. Again, the highest AIV (i.e. the greatest turnover of the investment in one year) is the one to consider. AIV shows you - and him - how hard your - or his - money is working for you.

THE NEW TERMINOLOGY

Sharp-end advisers such as myself – I don't like the overworked 'consultant' word – succeed if our clients make more profit. It can be hard work at the sharp end to coax even a reasonable bottom line for my pig clients; and it is the ones in trouble who seek help, of course. This is true anywhere in the world, in fact.

Why the need for new yardsticks?

There is nothing radically wrong with the old terminology. After 70 years of use, it is certainly very familiar! Even so it is not good enough for today's conditions. We can do better.

The existing terminology is based largely on *performance*. Today *profit* matters on pig farms far more than it ever did. You can have very good performance but still make less profit (Table 1).

Table 1. PROFIT RATHER THAN PERFORMANCE

Profit	Here are some actual returns from people in each category. Sale liveweight av. 87 kg			
(n)	*Pigs Weaned Sow/Yr (Pigs sold Sow/Yr)*	*Wt of saleable meat per sow/year (kg)*	*Wt of saleable meat/tonne feed (kg)*	*Relative Nett Profit per pig/sold %*
2	29.7 (26.9)	1778	229	109%
8	27.1 (25.7)	1642	235	111%
14	25.0 (24.1)	1540	240	115%
100's	22.0 (20.1)	1285	209	100%

Comment : **The guys at 25 pigs weaned/sow/year made most profit.** Note how meat produced per tonne of feed is a more reliable guide to profit than meat produced per sow per year.

Source: Clients' records

SLC & MSC

Profit from animal production comes in two main practical ways. Either produce more at the same cost (MSC) or produce the same at less cost (SLC). There is a third way, of course, to produce more at less cost, but sadly this is beyond the reach of most cash-strapped pig producers in an end-market. Few of them are able to control pig price – or even to influence it. Pig price is volatile, and will be for many years to come.

So if everyone produces more at the same cost, which is possible, overproduction results and the pig price collapses. However, if we all produce the same but at a lower cost, which is also possible, the market is not overloaded, the pig price remains firm and the cost savings can be pocketed as profit.

The immediate question which stems from this argument is 'How good must performance be to generate an adequate return?' The answer varies on each farm, but is closely linked to profit which can be forecasted within reason if the producer is skilled at it. In fact projected profit, not necessarily forecasted or targeted performance is the key to success. Figure 1 gives the type of matrix I use with each client to get the balance right. And it usually results in a better bottom line than merely using the [currently-favoured better performance/higher output concept as his primary goal.

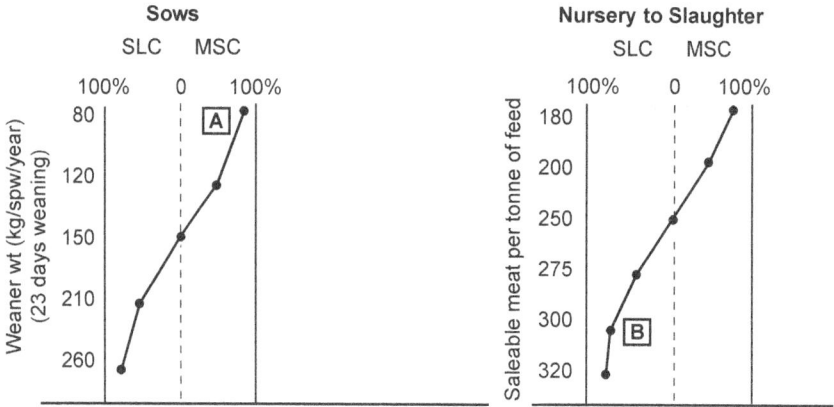

Figure 1. SLC or MSC: Where to put the effort

Saleable meat = dressed carcase weight – what the pig producer is paid for.
SLC = producing the Same at Less Cost
MSC = producing More at Same Cost

How to use this table
Calculate the amount of weaner weight produced per sow per year in kg; and the amount of saleable meat (kg dressed carcase weight) produced per tonne of feed from exit from the nursery to slaughter.

Read off on the graphs how much attention (in %) is needed to devote to SLC or MSC in proportion. For example, if at **(A)** the producer needs to devote about 60% of his time to improving sow/weaner performance and perhaps 40% to saving costs without destroying what breeding performance he already has. At **(B)** however, the physical performance (FCR:ADG) is so good that improvement is unlikely to occur without incurring high costs. So while trying to maintain this performance devote about 85% of the time towards reducing the costs of doing this.

The key (left-hand) performance scales vary between national pig industries, of course. USA would have very different reference scales to Britain for example, which is why they should be constructed from local national figures. If 5 reference points are taken for each scale, the sigmoid shape will nearly always emerge, some more pronounced than others as in the two examples given.

Data from MLC Yearbook 2001.

Are present nutritional terms adequate?

I believe not. Take FCR for example. A useful yardstick *if it is used properly*. This is difficult to do on a busy working farm. Only 1 in 5 pig producers attempt it, and 18 years as a trouble-shooter in the feed trade showed me that pig producers still get it wrong by 0.2:1, which is equivalent to a rise or fall in their price per tonne of feed of 15%. Sure, researchers and academics can still use FCR as they *can* measure it accurately. But farmers need something better, more useful and more pragmatic than FCR.

Abandon FCR? Heresy! No, it is *not*. MTF is better!

Meat per Tonne of Feed (MTF)

This is a much better terms for us all to use. It sets the amount of saleable meat (in terms of dressed carcase weight) against the amount of food used to obtain it. Easier to measure than FCR, no weighing of pigs is involved (a hated task) as the producer gets his saleable meat figure from his processor's return. No weighing of food is involved either as a rolled-over figure for a week's or month's use can be obtained form the feed delivery dockets or invoices which are set against output over a period of his choice.

I've measured the rollover system's variation against precise batch-on-batch measurement and, if it is done diligently, the variation is ±2%, quite realistic and useable in farm practice. MTF is certainly better than FCR alone as on a typical, well-run farm the likely FCR error is nearer seven times as much, so I've found after measuring things carefully.

MTF is a good profit yardstick, as it places 100% of the producer's income against about 60% of his total costs – i.e. feed cost. Not ideal – nothing is – but it is still better than FCR. And from a feed manufacturer's view in particular it sets 100% of their customer's income against the cost of 100% of the product they are primarily concerned with – feed. Which is ideal for them. So why do so few of them use MTF and remain wedded to the FCR concept?

After all, if the producer has a good MTF figure then the FCR figure will also be good. Why? Because meat is 65% water, and if more meat is produced per tonne of food, then less food has been used to produce one kilogramme. Thus MTF parallels FCR to a large extent without the hassle and with less inaccuracy on a typical pig farm.

Table 2 illustrates how simple it is for the customer – or feed salesman – to calculate MTF. If sums are unfrightening, as they are in this case, people will use them. Table 3 gives the MTFs I come across worldwide – some surprisingly poor.

Table 2. HOW TO CALCULATE AN MTF FIGURE

(1)	Establish how many pigs are produced per tonne of feed *e.g.* FOOD EATEN 250 kg $$\frac{1000 \text{ kg}}{250 \text{ kg}} = 4 \text{ pigs/ tonne}$$
(2)	Calculate saleable meat produced / pig *e.g.* 75 kg liveweight increase x 75% killing-out percent = 56.25 kg (deadweight per pig)
(3)	MTF = 4 pigs x 56.25 kg = 225 kg M̲eat per T̲onne F̲eed

What is a good MTF figure ?

Table 3 suggests what these could be on a world basis. Because of this, note the modest FCRs quoted; even so they are fairly typical of average producers worldwide. (The FCRs you see bandied about are usually optimistic, I find, once I measured things carefully.)

Table 3. ACTUAL PERFORMANCES, WORLD-WIDE AND UK 2000 (25-100 kg)

	← World-wide→			← UK →	
	Very Poor	*Typical*	*Good*	*Target*	*Exceptional*
ADG (grammes)	518	617	681	800	1000
FCR	4.00	3.50	3.25	2.60	2.28
Extrapolated from these figures, at 55 kg of saleable meat per pig (73.5 K.O%) ...					
Saleable meat per tonne of feed (kg) 183	209	249	281	321

It seems that, at present, across the world a figure of 250-275 kg of saleable meat per tonne of feed is the one to achieve. However, in certain countries, and among certain producers, performances are higher than these world averages and 300 kg MTF is their current target with the top 5% producers achieving 325 kg MTF All these refer to the 25-100 kg weight range. UK figures incorporate 50% entires.

Price per tonne equivalent (PPTE)

Now to another new term.

We all know that cost/tonne can be a fickle, misleading jade. Even so, producers are still hooked on it as a primary buying – or refusal-to-buy – motive! PPTE is an excellent way to sell feed quality when good quality inescapably costs more per tonne, and so sets price resistance in motion. PPTE is another very simple ongoing calculation following on from MTF. For example, if the feed trial shows that feed A (costing £7 ($13; €10)/tonne more) produced, say, 15 kg more MTF than food B, then with a current dcw pig price of, say, £1/kg, expensive food A is therefore worth 15 x £1 = £15 ($27; €22)/tonne *more* than the £7 ($13; €10)/tonne cheaper food B.

This puts an 'expensive' food in a better and easily-understood light.

Feed compounders, please note! Why aren't you using MTF and PPTE to further your sales approach?

A WORD OF CAUTION

Just like the old term FCR, the benchmark value of MTF depends on the weight range involved, as pigs are much more efficient at turning food into meat early on in their lives than when closer to slaughter. So MTF should always be expressed as, for example "A target of 275 kg MTF across the 25-100 kg weight range" written as '275 kg MTF (25-100 kg).'

Return to Extra Outlay Ratio (REO)

This, I find, so useful! Liquid capital in pig production is available, but is sometimes limited and can be expensive. The choice of adding value to feed is very wide – I know of 67 feed additives or alternative dietary products which are on offer in Europe alone, and there must be more – probably nearer 100 worldwide?. Generally a producer seems willing to invest another 8 to 10% of dietary cost to protect, enhance or extend nutritive value or help lessen disease. Marketers call it 'adding value'. But which additive to choose?

REO calculations help him rank what products are under consideration or which are being sold by persuasive selling, as it shows what **Return** is likely from the **Extra Outlay**. Some additives cost more/tonne, some a lot less. Based on the published performance improvement claims (and their veracity has to be up to the buyer to decide) REO makes a value for money comparison easier as it compares input costs to likely return on a per tonne basis. REO is not the same as ROC or ROI; these latter generally refer to major capital projects, while REO is all to do with added value and therefore is especially useful in feed manufacture and choice of feed supplements. Table 4 gives an example.

The producer has a choice of investing the amount earmarked for added value on one or perhaps two major alternatives/additions, which from published evidence is likely to give a worthwhile payback. Alternatively, he could invest in, say, three or more additives which give additively the same return but have a much lower inclusion cost/tonne.

Table 4   HOW REO CAN BE USED TO COMPARE POTENTIAL FEED ENHANCER ADDITIVES Conversion rates: £1= $1.82, €1.47

Total dietary cost £160 ($290, €235)/tonne. One major dietary enhancer costs 10% of the dietary cost (£16/tonne) and yields an REO of 3:1 *£16 x 3 = £48 ($87, €70)/ tonne. But… let's look at 3 other product options.

	Cost per tonne			*Expected return per tonne*		
Additive A	1.0%	=	£1.50	REO	8:1	=	£12.80
Additive B	2.0%	=	£3.20	REO	10:1	=	£32.00
Additive C	0.5%	=	£0.80	REO	5:1	=	£4.00
		($10; €8) £5.50				($87; €70) £48.00

If both options give the same very good payback – why bother to change? Good question! *Because the three alternative options only cost £5.50 tonne in place of £16*, so the £10.50 ($19.10; €15.40)saved for the same projected benefit can then be directly added to the profit pool, (or re-invested in further potential dietary benefits). See what I'm getting at? This is an example of the *same for less cost* (SLC) strategy discussed earlier.

REO is a very useful diagnostic tool.

Generally, the products with the highest REOs are the ones to use first, for example the trace element proteinates (a trace element linked to an amino-acid) can yield REOs of well over 20:1. And that's using capital well!

Annual Investment Value (AIV)

Obtaining a good REO is all very well, but supposing it takes an age to secure? A banker, when considering a loan, looks for how long the investment might reasonably take to pay back the sum advanced out of extra income – as well as his usual caution over credit-worthiness. For example, the REO potential on a creep feed additive is 21 (365 days ÷ about 17 days of use). On a lactation feed 17 (52 weeks ÷ 3 weeks of use); a weaner feed (on an 8 week nursery turnround) 6½ times; a grower/finisher food 3 times and a gestation diet only 2¼ times.

In other words, how hard is the money working for you? AIV tells you this.

So AIV, which is another simple calculation, reflects the unit value of the additive multiplied by the annual turnround. Let's take one of dozens of examples where the REOs at first glance may not be as attractive an investment as it seems. Thus a lactation feed additive with a modest REO of 2.0:1 has an AIV of 2 x 17 = 34 – an excellent use of money – while a growth enhancer with a *more attractive* 4:1 REO nevertheless only has an AIV of 4 x 3 = 12. AIV shows you where the money works hardest.

Here are two more simple terms we are all guilty of using but which are misleading and out of date. They always were, so let's bin them.

Pigs per sow per year (18? 22? 25? Etc)

What a silly yardstick! Why? Because it takes no account of the amount of *weaner weight* with which that poor old (over-burdened) sow gave you. You need to know this. Table 5 shows how misleading the term can be.

Table 5. WEIGHT OF WEANERS PRODUCED PER SOW PER YEAR (kg)

	Weaning age (days)	Poor	Typical	Good	Target	Exceptional
SEW	10-12*	n/a	n/a	103	103	n/a
Conventional	21	81	89	133	147	202
Conventional later weaning	28	97	109	184	217†	[—]
Swedish conditions	35	115	133	261†	299†	[—]

* Insufficient data for SEW technique. [ ] Exceptional producers outside U.K. rarely wean over 24 days.
† These sorts of productivity place a considerable strain on the sow thus can be difficult to maintain consistently.
All figures are corrected for farrowing index.

We need to replace it with…..

Weaner Weight per Sow per Year (WWSY)

Weight for age at weaning (wherever weaning is) has an important effect on subsequent economic performance to slaughter, so WWSY (Weaner *Weight* produced per Sow per Year) is a better indicator than just 'pigs' or even 'weaners' per sow per year, providing the date/time of weaning is attached to it. The new measurement should be written as…

158 kg WWSY[24] (158 kg of weaners produced per sow per year weaning at 24 days).

Next, lets discard the old favourite…% mortality.

% Mortality (to weaning is not an efficient yardstick)

Why? Because it is misleading (*Table 6*)

Table 6. WHY '% MORTALITY' IS AN INEFFICIENT YARDSTICK

Looks bad –	15% mortality of 12 born-alives is 10.2 reared
	$(12 - 1.8 = 10.2)$
while…	
Looks good –	5% mortality of 10.75 born alives is also 10.2 reared
	$(10.75 - 0.54 = 10.2)$!

We need to replace it with an…..

Absolute Mortality Figure (AMF)

How many pigs died per litter of those born alive? Not the percentage which died, *as we don't know the litter size when just a percentage is quoted*. This is because the more piglets which get born alive, so the relative percentage mortality will increase.

Express it as AMF 1.2 of 12 b/a (b/a = born alive), or AMF 1.2/12 (ie 10% mortality).

However AMF 0.8/12 is only 6.66% mortality – good. While AMF 0.8/8 is 10% mortality again, not so good.

Look at Table 6 again which shows you that in expressing it like this, the *problem is the born-alives*, not the mortality! Once you see the logic, you'll get your head around it!

These are just six of twelve new terms I am using to guide both the pig producer – and myself – towards a more viable business through profit-orientated rather than the performance-driven measurements most people use today. Maybe you should consider the concept? I hope so.

NEGOTIATING SKILLS

I published this short article in 2003. Since then I have received more emails, letters and phone calls praising its logic than for any article since I won (to my considerable surprise) the national "Business Writer of the Year Award" in 1996. I think its impact comes from the errors in negotiating I made myself over a period of many years. There is nothing more rewarding than learning from your own mistakes!

By the way – I still talk too much!

Since my recent textbook on 'Solving Pig Problems' was published, several readers have written to me saying why didn't I cover 'negotiating' in the 96 page Business Section? So... tripped up by my own title for this column, here is a piece on 'What My Own Textbook Didn't Tell You'! I expect there will be more omissions, and this column is an ideal place for them. (Lack of space kept 'Negotiating'; and some other subjects out of the book; after all, 590 pages is enough for any textbook!).

We've all been negotiating since we were babies, trying to wheedle our parents, grandparents, brothers and sisters to give us what we wanted, or not have imposed on us what we didn't like!

Since then we have negotiated our way through life with varying degrees of success. Years ago I put myself on a negotiating course – one of the best things I've done because I not only learned a lot, but had to *unlearn* even more!

What do I mean by this?

Curb those negotiating instincts

- Don't talk too much. The more you talk the more you are likely to give away in the end. Encourage the other fellow to talk, he may reveal something about his position. The key to communicating is listening. Sounds odd, I know, but try it and you will see how true this is.

- In the same vein, do not give away your position at the start of the negotiation. Come to it gradually through discussion and questioning. Farmers are too forthright, impatient, anxious to get to the nitty-gritty.

- Instead, start assessing the other side's expectations as soon as you can. This will give you an idea of what concessions you may need to make. Sometimes this will tell you that there is no need for any at all!

What to give away

- With regard to concessions, prepare in advance and have clear in your mind which concessions will cost you little, but could be worth a lot to the other side. A good appreciation of the market you are both in is always homework worth doing, especially in the other side's field of expertise.

- If you are likely to be asked for a price reduction, have a list of questions to ask to find out if the other side's real concern is something else. Poor sales, cash flow, uneven deliveries, unexpected expenses, labour costs, transport costs? Discussed sympathetically, but not intrusively, this can deflect or reduce the demand. Can you help him with any of these?

- Do not give up anything without getting something in return.

- Negotiating with farm staff is rather different from negotiating business-to-business. Staff negotiations tend to be more formal and involve the outcome you **both** want, as the relationships are on-going and not concluded at the end of a deal, when the business negotiator disappears to the end of an email.

- Leave haggling to the concluding moments, not early on. Haggling is for the end-game when you have to bridge a final gap. Do it too soon and you squander the fruits of skilled negotiation. Haggling is the antithesis of negotiation. Farmers love to haggle!

- Similarly, don't go for broke too soon. If you need to buy 100 units ask the price for 50 first, which puts you in a position to negotiate a discount on 100.

- Preferably, don't negotiate with Managing Directors! They are usually too inflexible, or proud, or self-important. Sales managers and buyers are better bets in my experience.

- Finally, agree things *in writing* after the handshake.

One final observation. Good negotiators tell me – and I have had it done to me – that there are two key words which usually help construct a satisfactory deal. 'If' and 'then'. "*If* I do this, *then* can you do that?" That little, subtle approach when stalemate looms can make all the difference.

Try it, it works!

Managing to maximise profit

.

THE COST OF POOR GROWTH RATE

Generally textbooks cover growth rate well, but there are aspects such as the economics of slow growth, which none of the books on pigs I've read – and I have most of them on my shelves – cover adequately enough. So let us look at costs in particular, which I guess must top the list of omissions.

Growth rate, as we all know, is an important measurement, as fast growth to slaughter generally uses less food – a slaughter pig finishing a week faster saves 7 days' food – and at its maximum feed intake too – from its total feed requirement. This one week's faster growth is equal to, on average, an increase in pig price of 5p ($0.09, €0.07)/kg or 5% at the time of writing.

Generally speaking you cannot grow a pig too fast to 30-35 kg because of the young pig's superior food-converting ability. Beyond that initial acceleration-phase, growth rate has to be balanced with food conversion and carcase quality (grading) so as to maximise income and keep costs to a minimum.

Cost-driven

"Keeping costs to a minimum". Well now, that's true enough! The main problem in most European pig countries is that our current costs of producing all the food we eat are 20% too high on a global basis, and if our pig industries are to progress in a world market on a long-term basis, that is the sort of target cost reduction we Europeans have to aim for. Our current superior productivity is only temporary; this is steadily being eroded by competitors outside Europe who are rapidly increasing their use of modern technology available to everybody, *because they are investing more*, and more rapidly than some of us, in housing, genetics, labour and records.

How outdated performance terms stultify investment

As I've mentioned elsewhere in this book, most textbooks correctly and accurately quantify the advantage which quicker growth rate can provide but they never express it in MTF terms (Extra Saleable Meat per Tonne of Feed used). An omission, because we sell *meat* (not pigs), and if faster growth sells more meat for us for the same or similar costs – which it does – then surely it should be mentioned in the same breath as other performance achievements like FCR, DLWG and Yield (KO%) (*Table 1*).

Table 1. ACTUAL PERFORMANCE WORLD-WIDE 1996-2003 (25-100 kg)

	Typical	Good	Target	[Exceptional*]
A.D.G. (g)	617	681	763	900
At 73.5% KO% ... Saleable Meat (kg) per Tonne of Feed (M.T.F.)	220	245	275	324

* Achieved now only a few farms, this figure gives some idea of what targets will be in the next 5 to 7 years.

Table 1 shows the differences in saleable meat due to slower growth which I meet every day on the farms of typical producers compared to the best I visit. It is a massive 55 kg less saleable meat for every tonne of feed fed between 25-100 kg. And meat generates around £1 sterling ($1.82, €1.47) for every kg shipped while food costs only a seventh to a tenth of that!

Why express it this way? On a tonne of feed basis, not the F.C.R. or A.D.G. terms beloved of the textbook writers?

Because the savings can quickly and easily be related to the three economic areas which primarily affect growth rate ...

Food quality and cost : The cost of minimising *Disease* : The cost of a *correct growth-enhancing environment* (especially ventilation).

After a good pig price, which we can only influence marginally, life today is all about minimising costs and re-investing wisely.

Looking at reducing costs in a new light : how it stimulates investment

FEED COST

Taking the 55 kg deficiency in meat sold between the typical and target growth rates today, at a very modest deadweight price of only 100p/kg ($1.82, €1.47/kg), producing 55 kg more meat x £1 = £55 ($100, €80.1) per tonne, is therefore equivalent to a feed reduction of say 33% on all the grower/finisher food you will buy in future.

This huge figure is a real incentive to the producer to invest in better feed quality, raw material analysis, new feeding systems like wet (pipeline) feeding and recent techniques like multiphase feeding and challenge feeding (to establish current immune demands), any one of which can contribute materially to achieving that 55 kg goal.

DISEASE COSTS

Using that £55 ($100, €80)/tonne analogy, at an achievable 200 kg food/pig consumed from 25-100 kg (*ie* a FCR of 2.7:1) this is 5 live pigs/tonne feed, or £11 ($20, €16)/pig (€15.50) increased income. A farmer producing a modest 1000 pigs/year should be employing a veterinarian to disease-profile and thus safeguard this valuable potential output with accompanying medication/vaccination. Yet most don't! The £55/tonne potential benefit translates into a massive £11,000/ year ($20,000, €16,000) investment on such a small herd, and the extra cost of disease profiling is nowhere near that (*Table 2*).

Table 2. BEFORE-AND-AFTER RESULTS FROM USING A PIG SPECIALIST VETERINARIAN TO DISEASE-PROFILE 3 FARMS, WITH EXTRA VACCINATION & RE-MODELLING EXPENSES COSTED IN. (US$ PER SOW)

	Before			*After*		
Farm	*A*	*B*	*C*	*A*	*B*	*C*
Estimated cost of disease per year*	284	186	300	80	96	109
Cost of veterinarian	8	3	12	30	27	31
Cost of vaccines & medication†	26	18	30	18	20	21
Cost of remodelling (over 7 years)	–	–	–	27	45	33
Total Disease Costs (US$)	318	207	342	155	188	194
Difference (Improvement %)	–	–	–	51%	9%	43%

*Disease costs *estimated* from items like the effect of post weaning scour and check to growth on potential performance; respiratory disorders, ileitis, abortions, infectious infertility, etc.
†Note that the cost of planned preventive medication was *lower* than for reactive curative medicine.
Source : Clients' records and one veterinary practice

From the costings in Table 2 the average cost of disease profiling *and* the farm remodelling costs recommended by the veterinarians concerned was still only 23% of the animal disease cost before disease profiling by the vet was introduced. Veterinary services alone, only 8.7%. Both figures are way below the nominal

£11,000 ($20,000, €15,600) difference suffered by the typical growth rate achieved in Europe today compared to the possible target.

ENVIRONMENT COSTS

Or you may care to spend the £55 ($100, €80)/tonne or £11,000 year or £11 ($20, €16)/pig on updating the ventilation pattern, or insulation, or whatever. Both are dauntingly expensive, but not when looked at over a depreciation (amortization) period of say 7 years, which is as long as most good ventilation/insulation systems should last. During those 7 years our 1,000 house piggery suitably updated should generate an extra £70,000 ($127,400, €102,900), more than enough to pay for the update/renovation plus accrued interest. About 3 times more, in fact, even at today's expensive prices for state-of-the-art housing modernisation.

Final questions

At this stage the producer, cautious as ever, says "Yes, but *will* the improvement you've cited be realized?" In my on-farm experience, almost always. Especially with good guidance. Moving from typical to target growth rates is surprisingly achievable these days – most pigs are underachieving by 90g/day! That's a 14 day reduction . You see, we have the technology available to us, the problem is that many pig producers are slow to adopt it on cost grounds – back to costs again!

In other words, lack of updating investment. This is the *true* 'British Disease' – from Governments, businesses to private individuals. Even I invest 15% of my yearly income after tax on being trained, and still wonder if it is enough?

Overheads contribution ignored

My last plea is for the textbooks to include nominal overheads in their growth rate calculations. They rarely do, and neither do the academics in their research reports. You and I know that overheads are a growing burden on costs, in particular good labour and housing costs and sensible preventive medication costs are all escalating rapidly these days wherever we farm – not just in Europe. The experts must get with it! Table 3 illustrates this. *Saved overheads can add up to a further 40% to any performance-related savings from FCR and DLWG alone,* and 30% is quite common among my pretty efficient clients. Not to include them reduces the likelihood of the producer investing in the very things he really ought to be doing to ensure he achieves the improved performance potential I've quoted!

Table 3. OVERHEADS ARE OFTEN OVERLOOKED WHEN EXAMINING FASTER GROWTH

Assumptions:	Pigs 30-100 kg. Dressing % 73%. Av. food cost £140/tonne ($255, €206). Av. overhead costs/day (including capital depreciation) 12 to 14p ($0.22 to $0.25, €0.18 to €0.21)

Days to slaughter	Food eaten (kg)	Overall FCR	Overall ADG (g)	Food cost/ pig			Overheads cost/pig		
100	210	3.0	700	£29.40	$53.50	€43.20	£14.00	$25.50	€20.60
90	189	2.7	777	£26.46	$48.15	€38.90	£12.60	$22.93	€18.52
80	168	2.4	875	£23.52	$42.80	€34.57	£11.20	$20.38	€16.46

Savings per pig from improved FCR vis-à-vis lower overheads:–

Days to slaughter	Savings in food/pig			Savings in overheads/pig			% of overheads compared to improved FCE
100	–			–			
90	£2.94	$5.35	€4.32	£1.40	$2.55	€2.06	47.6%
80	£5.88	$10.70	€8.64	£2.80	$5.10	€4.12	

I said earlier that many European producers have to get their costs down by 20% to compete with the new global players. This article suggests that by increasing growth rate from weaning to slaughter by 7 days will achieve a quarter of this target.

Double it (by an eminently achievable 90g/day on many farms), and you are half way towards a more secure future!

Deamination and growth rate

Pig producers tend not to understand this phenomenon. I'm often asked, in a puzzled way, "Why, when we boosted the protein level in my weaner feed, did the pigs relish the food but grew noticeably slower?"

When supplying the nutrient specification of a pig diet from a particular range of ingredients, specific amino-acids are oversupplied as the minimum requirement of the first limiting amino-acid is met. With diets formulated from traditional feeds, lysine is the first limiting amino-acid and hence when its requirement is met all other amino-acids are oversupplied and are wasted to the extent that they cannot be used for tissue synthesis. Worse still, the excess amino-acids are broken down (deaminated) and the nitrogen is excreted in the urine. This is an energy-demanding process and for each gram of protein oxidized 4.9 kilojoules of energy is lost in the urine and a further 6.6 kilojoules is lost as heat. Thus presenting the pig with an oversupply of digestible amino-acids results in a decrease in energy utilisation of 11.5 kilojoules per gram of excess digestible crude protein. Without that energy available for productive purposes, the pig grows slower.

So… when buying new, fast-track genetics, don't fiddle with (which usually means raising) the protein specs of the early grower food. Let a nutritionist do it for you. This deamination effect is seen quite commonly in home-mixed diets, and causes much head-scratching until it is explained, and then the diet adjusted by a nutritionist working with the breeding company or supplier of the AI semen.

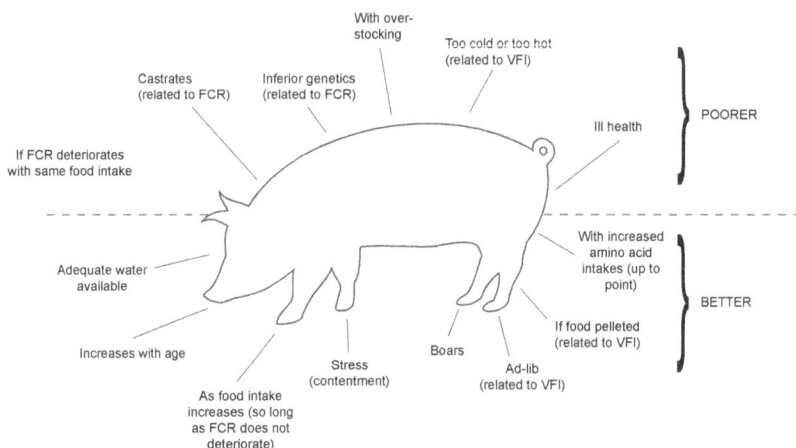

Figure 1. Factors affecting growth rate.

As we sell meat, not pigs, lean growth is vital to us. Many years ago Prof Colin Whittemore coined the phrases "The Acceleration Phase" and "Cruising Phase", of lean deposition. They have never been bettered. Figure 2 illustrates the concept in its most simplified form.

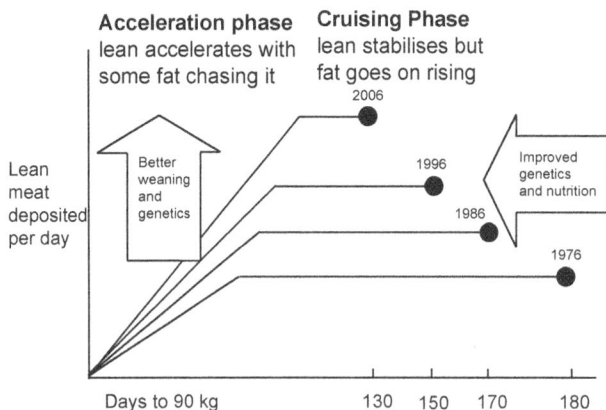

Figure 2. Growth rate across the decades.

WEANING WEIGHT

When I was farming for the late David Taylor (we had 1200 sows) we found that evenness of individuals in each litter was important (Table 1). In fact David always said that in profit terms the smallest pig in the litter cancelled out what margin one could expect from the biggest one (Figure 1) –advice which I have found to be not far off the mark in the subsequent 30 years that I have been involved in pig production across the world, and which Table 1 tends to confirm, I guess.

Textbooks do mention weaning weight, of course, but those in my library don't deal with it in detail, presumably because weaning weight variability has not been a priority area in research,obviously due to management variables clouding the issue, and hard evidence seems to be lacking. Perhaps I can pull a few threads together from my clients' experience.

What affects evenness at weaning?

GOOD INDIVIDUAL BIRTHWEIGHTS

Many people rely on *average* birthweights as a yardstick. This to my mind is dangerous. For example in one comparison a slight difference in average birthweights of only 8% (1.48 kg v. 1.37 kg) – which is not detectable to the stockman's eye – resulted in nearly twice as many piglets under 1 kg and 50% more over 1.75 kg.[1]

The bigger the pigs are when born the more will survive to weaning. They will withstand weaning better, and grow out faster and more efficiently. Overall the weaning weights will be less variable.

---

[1] Wells, D, *et al*. "New blood in our pig unit" (Easton Lodge). (UK) Farmer's Weekly 25 May 2001 p87.

So how can we influence weaning weight variability?

1. **Don't wean too early.** I have always been uncertain about the American desire to wean at 16 days or under. Except under the best, well-organised pig flow and housing management, this strategy piles up a heap of problems, not always counteracted on many of the 'industrial' US farms I have visited (the very reason why I was asked to visit, in fact!).

    The current work done by SCA (Dr Varley) in Europe in advance of the proposed EU move from 21 to 28 days mandatory weaning threshold is interesting. Their work shows distinct performance and even economic benefits from later weaning.

2. **Consider weaning by *weight*, not by *date*.** Here no pig is weaned under an agreed target weight. This used to be 5 kg but has now advanced to 6 or even 7 kg. Lightweights are moved to a nurse sow, or alternatively the heavier ones are, leaving the lightweights longer suckling access to their natural dam. This is called split weaning. It needs care and planning to do and, like cross fostering, may inflame a PMWS situation. Take advice on it, both from your vet and from a manager who has a successful track record on it.

3. **Split nursing.** Similar to split weaning, except that the extra access to milk is provided within the first 24 hours of birth. The best pigs in the newborn litter are removed to a warm clean and cosy location (NOT mixed with other litters) for about 2 hours. This allows the smaller pigs to get a really good feed or two, not necessarily of early colostrum, though this may well be involved within 12-15 hours from birth, but of rich milk at least. Said to be good for PMWS mitigation by the way, if initial comments are confirmed. Tokach (2002)[2] reports fewer lightweight pigs.

4. **Watch your sow lactation feeding.** This is not so much the sow 'nose-dive' in condition resulting from a variety of poor management stockmanship and feeding errors/omissions, but not getting as much of a special lactation diet into the sow as you can.

    Again there are a host of do's and don'ts here – from undue worrying about early lactation and udder problems, to temperature, water adequacy (how *easy* it is for the sow to drink as much as flow rate etc), prelactation

---

[2] Tokach, MD, *et al.* 'Dealing with Weaning Weight Variation'. Procs. Saskatchewan Hog Congress, (Nov 2002) - supplementary paper.

nutrition, wet feeding, etc, etc. Study this important subject – I cover it in detail in my previous textbook.[3]

5. **Correct feeding and management in implantation.** Implantation? As early on as that? Yes, I believe so. The way the blastocysts are encouraged to 'plate out' on the wall of the womb (endometrium) quickly and evenly must have a result on more born and even birthweights too. 'Late' arrivals either cannot get a toe-hold at all or if they do succeed in establishing themselves have less area available to develop into embryos. Result, more big pigs, sure – but also more small 'overcrowdeds' and lower litter numbers from those that don't implant or get squeezed if they do manage it and risk being forced out eventually.

Table 1   THE SUBSEQUENT PERFORMANCE OF LITTERS WHERE THE WEANING WEIGHT SCATTER (<4 kg :> 5.25 kg) WAS CLOSE OR WIDE

	Narrow scatter (theoretically better)	Wide scatter (theoretically worse)
Number of litters	120	126
Average number born alive/litter	10.3	10.2
Average birthweight of born alives (kg)	1.42	1.41
Average weaning weight at 22 days (kg)	5.36	5.11
% weaners under 4 kg	15	22
% weaners over 5.25 kg	72	67
Pre-weaning mortality of born alives (%)	10.6	13.9
Average days to 102 kg from weaning	161	167
MTF[1] (kg)	268	259

[1]MTF = (Saleable) Meat per Tonne of Feed fed (to 102 kg)
Comment: These records suggest/confirm that...

*   Litters with more smaller pigs are likely to have higher mortalities, take longer to reach slaughter and yield less saleable (d.c.w.) meat for each tonne of grower/finisher food fed.

*   Interestingly the increased income over food costs (at a current pig price of €1.25 (£0.85, $1.55)/kg dwt per pig) in the narrow scatter group in this trial approximately equates to the current gross margin on one pig more or less, as David Taylor (see text) always claimed. (He said that the smallest pig in the litter cancels out the profit from the best one.)

    See diagram overleaf. . .

[3] Gadd, J. "Pig Production Problems - A Guide to Their Solutions" (600 pps). Nottingham University Press (2002). Textbooks available from orders@nup.com

Figure 1. David Taylor's firmly-held belief (from 26,000 farrowings on his Taymix farm) as to how *uneven litters* (despite *good litter numbers*) run away with profit.

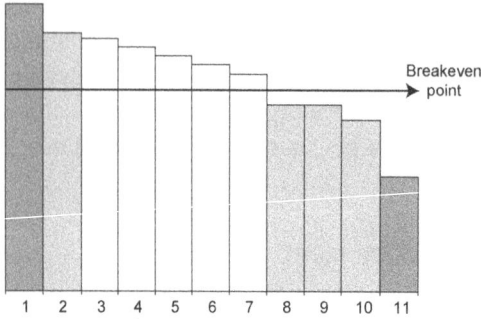

The Ideal .... an even litter

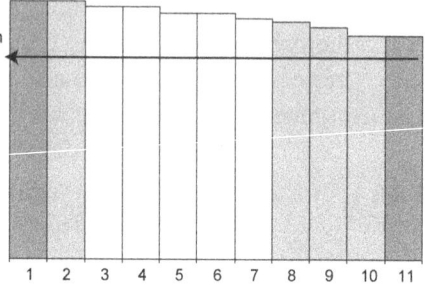

First: Pig no 11 in the litter (the smallest) cancels out the profit on the best

Then: Pigs nos 8, 9 and 10 together cancel out the profit on the second-best pig in the litter (no 2).

Maximum profit is now retained on pigs 1 and 2, and a decent profit on pigs 8 to 11 secured as well.

Two quotes

David Taylor, "An even litter is worth 40% more profit to me"
Dr Peter English, "Farm for the smallest pig in the litter".

... and later on, to slaughter?

Variation in shipping-out weight can cause big 'slowing-down' losses from increased overheads. 17% less gross margin is common. Minimize it by . . .

a)   Separating the sexes and feeding them differentially.

b)   Rearing the 15% lightest separately, once they leave the nursery.

c)   Split large groups into 'biggies' and 'smallies' as routine.

d)   Choice feed (still experimental) see page 251.

e)   'Stream' pigs affected/then cured, but watch housing costs. See page 241.

THE PROBLEM OF WASTED FOOD

Pig producers are not stupid, they don't waste food on purpose; what is happening is that they don't realise that so much is being wasted. In three ways:-

1. **Physical Wastage** (FCR +0.26): Trough spillage, going down the slats, being trodden uneaten into muck; rats, birds, mice, etc.

2. **Nutritional Wastage** (FCR +0.15): Feeding the wrong food, wrong feed-scale; to the wrong nutrient density pattern; offering stale food, contaminated food. Lack of water/water flow and poor access to water.

3. **Environmental Wastage** (FCR +0.25 in severe climatic conditions, +0.15 otherwise): Pigs too cold, too hot, too overcrowded, too much aggression at the trough or drinker; gases and dust; poorly designed pen layout, etc.

This adds up to at least 0.5 worse FCR on what is possible genetically! Some of you will think this an exaggeration. I can assure you it isn't! Do you wonder why some leading experts can get FCRs from 7 to 105 kg of 2.3:1? When many producers get nearer 3:1, or worse? These guys have really attacked food waste on all three fronts, I find, and it pays handsomely.

Direct food wastage

The average feed hopper dispenses 16 tonnes of feed each year. Even with pellets, wastage is rarely less than 5% and averages between 6% and 7%, and on some farms can be as high as 15% (*Table 1*). Wasting 6% of 16 tonnes/ year is not far short of one tonne of feed (960 kg), wasted from each hopper every year! It need only be 2% if the latest low-waste hopper designs (e.g. Big Dutchman) are used.

THIS IS HOW MUCH FOOD YOU ARE WASTING/YEAR

The average finishing pig producer wastes about 6% of food (ie food put in the hopper but not eaten). However 32% of the producers checked were wasting 10%, and 12% were wasting 15%.

Measurements were taken by placing 0.45 m x 1 m grilles under conventional feed hoppers set into a raised plinth. An allowance (deduction) of 10% extraneous collected material (faeces, dust, etc) was made. Pigs were grown from 20 kg to 88 kg converting at an average of 2.8:1.

Table 1. YEARLY FOOD WASTED – TONNES (1000 kg)

Pigs produced each year	*2%*	*6%*	*10%*	*15%*
500	1.9	5.7	9.5	14.25
1000	3.8	11.4	19.0	28.50
2500	9.5	28.5	47.5	71.25
5000	19.0	57.0	95.0	142.50
	This is what is likely with well-designed hoppers	This is typical today on many farms	32% of producers sampled	12% of producers sampled

Table 1 shows the effect of direct waste on your major input cost. Expressed in terms of saleable meat lost per tonne of feed, Table 2 shows the damage another way.

Table 2. A PRODUCER WITH 100 SOWS SELLING 2000 FINISHED PIGS/YEAR, EVEN IF HE ONLY REALISES 75 kg SALEABLE MEAT PER CARCASE, RISKS LOSING THE FOLLOWING TONS OF SALEABLE MEAT EVERY YEAR FROM ONE OR MORE OF THESE AREAS

Fault	*Average wastage* recorded*	*100 sow herd; tonnes of saleable meat foregone per year*
1. Direct spillage	6%	10.75
2. Vermin and birds	0.5%	0.90
3. Poorly designed feeders	Over 6%	10.75+
4. Feeding meal/mash rather than pellets	3%	5.38
5. Feeding dry meal rather than pipeline wet/slop	4%	7.12
6. Feeding dry meal rather than wet/dry in 'shelf' and 'plate' feeders	3%	5.38
7. Incorrect wet/dry feeder adjustment	4%	7.12
8. Poor water management	Up to 3%	Up to 5.38

Source: Various surveys & experiments (1980-2002)

* Direct Wastage – as compared with Indirect Wastage which is:-
   wrong feed, wrong feed scale, low bushel weight, overstocking, insufficient feed spaces, etc.

Look at every hopper in the piggery and remember that each one could be costing you well over half a tonne of wasted food a year! You *must* do something about it!

What to do about direct wastage

1. **Try not to feed dry food to pigs over 25 kg**, and then only use pellets. Remember that the extra cost of pelleting is largely repaid by the improved FCR from better-digestible nutrients after correct hot-processing into a pellet has taken place.

2. **There are now at least 4 designs of economically-designed dry feed dispensers** which waste much less than 6%. I have experience of two of them, and can vouch for their low-waste properties, as both reduced waste to around 4%, or 520 kg feed saved per hopper per year. On a payback basis they paid for themselves in 7 to 10 months, an excellent bargain.

3. **Always put a cover on feed hopper** to protect from vermin, insects and atmospheric contamination. Feed picks up taint quickly and this could reduce intake – a form of indirect wastage.

4. **Never, ever floor-feed pigs!** The wastage has been measured at around 12% in performance terms. Yes, pigs can eventually clean up, but the food eventually ingested seems less digestible (*Table 3*).

Table 3. FLOOR FEEDING CAN BE VERY EXPENSIVE

In this farm trial, the producer was not convinced that floor feeding was very expensive. So we split the house into floor-fed and dry-fed sides by just putting 'Lean Machines' into 50% of the floor-fed pens. Results after 4 batches were…

(Pigs 30-84 kg)	ADG (g)	FCR	Saleable meat/tonne fed (kg)	Value/year
Floor fed	623	2.87	396	£523 ($952, €769)
Hopper-fed	710	2.54	444	£586 ($1066, €861)

Result: In this properly-planned trial, transferring to hopper-fed pigs increased income per tonne of food by £63 ($115, €93) per tonne. Each floor-fed pen contained 20 pigs and the hopper pen contained 18 pigs on average. Each hopper dispensed an average of 11.4 tonnes in a year so the extra value of the hopper's use was 11.4 x 63 = £718 ($1306, €1055), some 3.6 times more than the capital installation and borrowed money costs of the hopper.

Payback: On this basis, an improvement of only 0.1 in FCR (or 5 kg less food needed per pig) would pay for converting to a low-waste hopper from floor feeding. This is equivalent to each pig wasting as little as 3.22% on floor feeding, well under the 12% wastage usually quoted for this system.

Source: Clients' records

5.   **Wet the feed.** From the wastage point of view, I don't mind if you use wet/dry feeders or fully-pipeline feed into an *ad lib* or pulse-fed trough, though full pipeline feeding has other technical advantages over wet/dry feeders which could be very important in future pig nutrition techniques. These are Multiphase Feeding; Choice (Menu-Feeding); Enzyme-liberated cheap foods – grass, foliage, brassicas; liquid chemical nutrients, etc. These do not 'waste' the uptake/digestion of nutrients to the same extent as our current feed strategies do.

6.   **Heighten your awareness of signs of waste.** I often notice signs of direct food waste which managers don't, merely because I'm not so familiar with the premises, the pigs or the feeding routine, so the errors tend to be more apparent to a new pair of eyes. Do a waste audit once a month at least, concentrating just on signs of wastage during one tour.

More care needed on wet-dry feeder design and operation

At the last count there were at least 37 brands of wet/dry feeders in the world. I examine them carefully at shows and Expos. Only a few pass my personal test!

Faults in wet/dry feeder design

•   Many have troughs/bowls which encourage spillage

•   Many have poorly-designed throat adjustment – and some no adjustment facility at all.

•   Some have the dispenser arrangement too close to the pig's saliva, causing blockage or loss of the dispensation accuracy.

•   Some have poorly-sited and badly-designed drinker positions.

•   Many are designed down to a price by saving metal/using existing prefabricated containers, like cylinders which are more difficult to design-in properly.

•   Very few incorporate the 'wing' or deep trough concept which has been shown to reduce aggressive incidents at the feeding point (Baxter, 1987), lower stress and so improve feed conversion, by keeping the pigs eating longer at each visit to the hopper.

Stockmanship faults also cause feed wastage

Managing a wet/dry feeder to minimise waste and maximise satisfactory feed intake is a highly skilled job which takes patience, frequent observation and the willingness to make frequent slight adjustments to the amount dispensed per 'nudge'.

Thus *pen design* is important so that the feeder is easily visible and the adjustment handle accessible (extensions can often be welded-on if required). At the same time the feeder needs to be on the fringe of the resting area and thus 'go with' the general flow of pig movement round the pen. A normal flow is resting, then feeding, drinking, elimination, socialisation and back to resting. Exceptions are in large groups of yarded growing pigs/weaners, etc, where the feeders are centrally placed in the exercise area, either back-to-back or along a mini-fence. In such a layout one enters the pens to check the pigs daily, so feeder observation is easy.

Frequent adjustment – so that there is no aggression, no queuing, good long periods at the hopper, no piling of dry food in the hopper bowl (or on the shelf if the shelf design is used), no throat or canister blockage, and no interruption of the drinker – takes time and patient, repetitive observation twice a day. A new pen of pigs' feeding pattern needs to be learned all over again depending on stocking density and the individual's speed of eating.

Walker and Morrow's work shows what can happen if this is done carelessly, where the dispensation adjustment doesn't satisfy the pig's eating speed.

Table 4. THE EFFECT OF FEEDER SETTINGS ON PIG PERFORMANCE

	Low	*Medium*	*High*
Food intake (kg/day)	1.97	2.14	2.21
Liveweight gain/day)	727	797	845
Food conversion efficiency of carcase gain (g/kg)	3.70	3.58	3.47
Backfat thickness at P_2 (mm)	10.6	11.1	12.1

From Walker & Morrow (1994)

Table 5. THE EFFECT OF FEEDER SETTINGS ON PIG BEHAVIOUR

	Low	*Medium*	*High*
N° of feeder entries/pig/24 hours	51.5	45.6	42.2
Feeding time/pig/24 hour (minutes)	110	78	87
Queuing incidents/pig/24 hour	70	45	26

Table 6. Incorrect feeder adjustment equals 14% more on your feed price!

	(a) Too little[1]	(b) Medium[1]	(c) Adequate[2]	Diff (a) to (c)
Days to 100 kg	96.3	87.8	82.8	– 13.5 days
Food eaten per pig (kg)	189.7	187.7	182.9	– 5 kg
Saleable meat per tonne of feed	270	279	288	+ 18 kg (+ 6.7%)

Pigs 30-100 kg	1.Feeders pre-set, not adjusted.
	2.Feeders adjusted according to consumption residues

Taking on the Irish work still further, Table 6 shows how costly is too-infrequent feeder throat adjustment. 18 kg less saleable meat/pig equates, at today's pigmeat and feed prices, a penalty of 14% on the feed price per tonne. And this doesn't take into account the 13 days extra overheads incurred equal to another 1.6% higher feed price, at least.

Many single space ad lib feeders are placed out of reach of the access passage. This means the technician has to climb into each pen every day to check the settings. He rarely does!

Sure – occasionally the feeder has to be positioned out of easy reach. If this is the case, a twist-grip differential linkage can be installed so that the throat setting can be altered from a distance, thus ensuring it is done on a daily basis if needed.

It is not perfect, and is difficult to install so that it works over a period of time, but I've seen examples with which the producer was satisfied.

Reference

Walker, N. Some Observations on Single Space Hopper Feeders for Finishing Pigs. N.A.C. Pig Unit Bulletin 1990. pps 4-8.

WATER WASTAGE

Much has been written about waste of food, far less about wasting water. I cottoned on to this 40 years ago when at the Taymix farm, in an attempt to clean the place up a bit after some massive pig movements, we got a real dressing down from owner David Taylor at the amount of water-augmented slurry his tanker staff had had to remove in a week! Ever the man to watch the pennies (and quite rightly too) he calculated that the cost of the extra journeys would pay my salary for two months, and he had a mind to deduct it! Exit one chastened manager!

Wasting water is also expensive, as these notes written in the 1990's reveal. The position is the same today – only the prices have gone up...!

Water wastage

Does waste of water matter? Seems not, at first, as water is cheap enough! It is because it is cheap that farmers tend to let the pigs splash it about. Water wastage does matter for several reasons.

In Europe, work by Prof. Brooks, Dr. Carpenter and others showed that while the provision of water itself only cost about 1-2p a finished pig, *removing it as slurry* cost around £1 - £1.50 ($1.82-2.73, €1.47-2.2) a pig! They estimated that at least 40% of this could quite easily be controlled, thus saving at least £1 ($1.82, €1.47) a pig on the operating cost.

Water wastage is common in part-solid floored farrowing pens, the solid part being under the udder. So you use all-wire farrowing floors? OK, but European welfarists may insist on the lying area being solid (maybe even bedded, shavings or chopped straw) in future - it is shaping up that way.

Anyway, figures here suggest that wet farrowing beds (due to drinker or trough spillage) may increase mastitis by 40% and mid-growth stage *E. coli* scours by 100%. The costs of this have been suggested at €0.75/piglet and €2.00 (£0.50-1.35, £0.90-2.50)/ piglet *for every pig* produced, including treatment.

The answer to water wastage

The answer to water waste is not to reduce drinker numbers or reduce flow rate.

There is now an increasing amount of trial evidence to show that type of drinker, drinker height and the way the pig stands when drinking materially reduces water waste. Not only this, when the researchers measure pig performance from low-waste drinking systems the performance is often improved, sometimes statistically significantly, suggesting that some other as yet unsuspected factors may be at work.

Let me give you some examples.

TYPE OF DRINKER

Research in the 1990's showed that the performance of pig was not affected by the type of 3 bite and 1 nipple drinkers used under ad-lib feeding conditions, but daily gains and feed intake (but not FCR) were significantly improved when water was supplied through one make of *bite* drinker under restricted-fed conditions. The nipple drinker significantly increased water use (including wastage) possibly because of easy lateral movement as shown in Figure 1, drawing (b). So don't use nipples except for baby pigs.

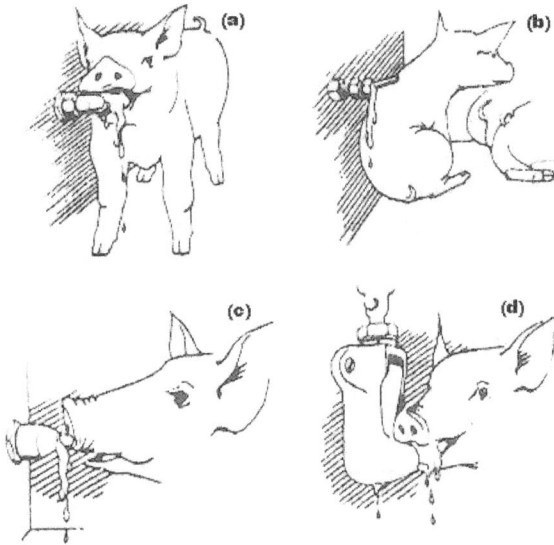

Figure 1. Water wastage by weaned piglets and growing pigs from the different types of drinker (Gill, 1989)

DRINKER LOCATION

Olsson (1983) found that pen hygiene was significantly improved and water waste reduced by 40% when drinkers were placed on the back of deliberate divisions between the sleeping and dunging areas (i.e. facing the dung area). 'Pigs seemed to be deterred from playing with the drinker as they had to make a circuitous route to get to it, thus they used it only when thirsty', thought Dr Olsson.

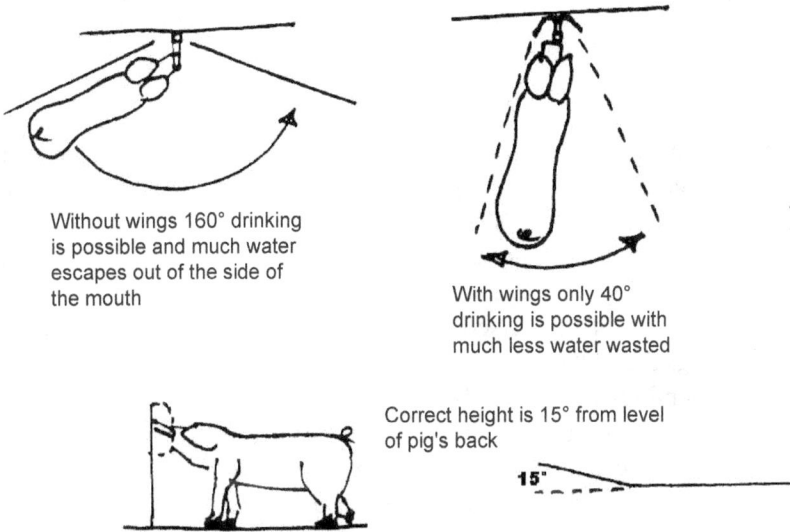

Without wings 160° drinking is possible and much water escapes out of the side of the mouth

With wings only 40° drinking is possible with much less water wasted

Correct height is 15° from level of pig's back

Figure 2. Drinker wings

This Dutch drinker-wing idea aligns the pig more towards a front-facing stance, thus the valve points down the pig's throat if set at the correct height (see below). I completed a farm trial which shows a benefit (*Table 1*).

Table 1.

	Water taken/day	Water wasted **	FCR *
Pens with no wings	5.8 litres	1.92 litres	2.72
Pens with wings ***	5.4 litres	0.86 litres	2.57

*FCR (25-84 kg) Pigs fed ad-lib
**Water collected from drain directly below drinkers
***Both treatments had 'Impex' height-adjustable drinker mountings set 15° above the backline of the pig.
Source: Gadd (1990)

Drinker wings are a pair of 'ears' or short 'wings' protruding at 45° either side of a bite or nipple drinker. They were originally introduced to protect growing pigs in particular from scratches and bruises from the projecting drinker and to prevent accidental operation of nipple drinkers, but it was soon found that the better alignment of the drinking pig wasted less water over a summer period at a flow rate standardised at 750 ml/min.

Considering these wings are so cheap and simple to make and fit, they should be fitted as standard procedure. Ask your supplier to make them; a simple enough metal pressing job.

DRINKER HEIGHT

Forget what many textbooks say! The correct height for a drinker is at a position which is most comfortable for the pig to drink. As the pig grows at about 4 cm every 16 days, no one height is ever ideal, so use:-

Spring-loaded height-adjustable mountings: These too make a major contribution to saving on water wastage. On the same client's farm we tried the Impex model against two fixed drinkers, one set at 300 mm and the other at 450 mm for pigs 30-62 kg. Again the benefit was noticeable (*Table 2*) when the adjustable mountings were altered once a week.

My advice is to fit spring-loaded height adjustable mountings as they markedly reduce water wastage, and so slurry production, up to one-third. In terms of slurry removal/slurry treatment costs, the payback is a few weeks only.

Table 2. VALUE OF SPRING-LOADED HEIGHT ADJUSTABLE DRINKER MOUNTINGS VERSUS TWO COMMON FIXED HEIGHT SETTINGS (30-62 kg).

	Water taken/day	*Water wasted/day*	*FCR*
Pens with Impex brackets	3.6 litres	0.7 litres	1.84
Pens with fixed brackets	3.7 litres	1.2 litres	1.91

* Pigs fed ad lib. 16 pigs per pen. (FCR differences are not significant)
Source: Gadd (1990)

The client has now fixed both wings and height-adjustable drinker mountings to the whole farm. He reports 2 tanker loads less slurry removed per month, a saving of 38%.

DRINKER ANGLE

Strangely an upward slope wastes less water – as researchers have found, but it also restricted water intake! Don't do it. Better is to present the angle at 15° to

20° from the horizontal as long as the height is correct (*see Figure 2*), which in simple terms is the tip of the drinker not lower than the pig's backline.

Follow these rules, use these simple devices, and you can use most makes of bite drinkers (for older pigs) with minimum waste - and maybe improved performance, as a pig which is comfortable at the drinker drinks more water, less stressfully, eats more food and so may perform better. Worth a try!

EQUALISING WATER PRESSURE

Studies conducted in recent years have shown that the rate at which water is delivered to the pig can have a marked effect on its performance. If water is delivered too fast, water is spilt making housing wet and cold and increasing the amount of effluent produced. If water is delivered too slowly pigs may eat less, thus grow more slowly and less efficiently (*Table 3*).

With conventional piping systems it is almost impossible to ensure that every pig gets its water at the correct rate to maximise its performance. While there are drinkers which can be adjusted quite quickly to increase or decrease flow-rate, it is a laborious task to calibrate each one, and so is almost never attempted, or if so, not maintained regularly thereafter.

Table 3. THE EFFECT OF FLOW RATE.

1. Young Pigs (10-25 kg)			*2. Sows*		
Flow rate ml/min 200	400	700	*Flow rate ml/min*	70	700
Daily gain (g) 210	237	249	Sow weight change in lactation (kg)	−14.8	−4.4
			Av weaning weight (kg)	5.89	5.65

Source: Brooks (1989)Source: Leibrandt (1989)
Note 2004:Flow rates for sows are now advised at < 1 litre/minute and the latest advice prefers troughs anyway (see page 156)

A patented watering system which did this at one calibration was called the Zeropipe. Once set, all the drinkers in the line delivered the same pressure and so flow-rate was standardised. This was a good idea to come out of Seale Hayne college (Dr Peter Brooks), but it sadly never really caught on.

Another advantage was that wide-bore plastic (pvc) piping was used (standard rain-water downpipe) so that all the water storage is in the actual circuit and is sealed away from dust, flies, urine contamination from rats (leptospirosis etc) and algal growth.

Badly-adjusted or leaking drinkers: These eventually get attended to, but often too late. The rate of water loss, and more important in cost terms, extra slurry removal, can be assessed as follows (*Tables 4 & 5*).

Table 4. LITRES WASTED IN ....

Stream size	4 hours	12 hours	24 hours	1 month
3 drips/second	6	12	25	750
Continuous 1 mm	19	56	112	3500
Continuous 2 mm	76	225	450	14,000
Continuous 4 mm	300	900	1800	55,000

Source: extrapolated from PIC, Canada (1988)

The average liquid slurry output of a pen of 10 pigs of 50 kg is 30 litres/day. On the above basis, failure to correct the leak increases the liquid slurry fraction you have to remove by the following (extra equivalent) number of pigs:-

Table 5. OVER A PERIOD OF ....

	4 hours	12 hours	24 hours	2 dqys
3 drips/second	2	4	8	16
Continuous 1 mm stream	6	19	37	75
Continuous 2 mm stream	25	75	150	300
Continuous 4 mm stream	100	300	600	1200

Source: extrapolated from PIC, Canada (1988)

Table 5 shows that if you keep 600 finishing pigs and one drinker leaks to the amount of a 2 mm stream for only 24 hours, you at once increase the slurry you have to remove by 25% or 450 litres, equivalent to the normal 24 hour waste liquid produced by 150 pigs.

Attention to leaks is a top priority on every farm.

What water wastage really costs

In Europe water provision has been measured, with the cost of removing it as slurry being €5.60 to €8.75 (£3.80-5.95, $6.90-10.83) per cubic metre (m³). The difference in water wasted between a well designed and sited drinker and a poor set-up from 11 to 110 kg is a staggering 364 litres! On current costings this is €2.50 (£1.70, $3.10)/pig.

Assuming two drinkers suffice 16 pigs in a pen with the pigs turned over 3½ times a year, choosing the right drinker and siting it correctly will save at least €140 (£95, $173)per pen each year, minimum.

Enough to make your eyes water!

MEDICATING PIGS
FEED OR WATER?

There are three ways of medicating pigs – by the feed, in their drinking water, and by injection.

For many years feed medication held the dominant position. Upjohn Pharmaceutical published Table 1 three years ago, and after I questioned pig specialist veterinarians recently, the opinion was that water medication has eroded 25% at least of the traditional preference for the feed route, as seen at the time of the survey.

About this time I published what I felt to be the pros and cons of all three options, and I update my findings here. I still remain ambivalent about the situation thus this section is written gathering in all the known claims and counterclaims so as to let you decide for yourself. No, I'm not opting out, but as I've said before in this book, after advising on hundreds of farms it soon becomes apparent that there are many differences between them and in the circumstances they face. These variations will encourage the producer and his veterinarian to lean one way towards food or the other towards water medication.

But if you push me, I feel the water route has it!

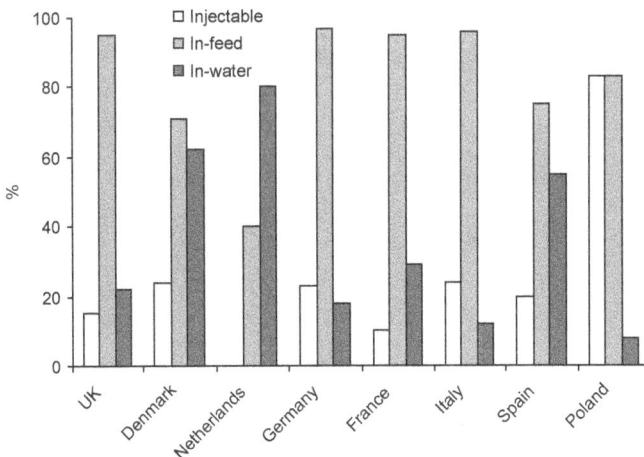

Figure 1. Preferences for delivery mechanisms of chemotherapeutics to commercial pig herds.

Water medication – advantages

- Generally speaking, sick pigs will drink when they are reluctant or unable to eat.

- Sick pigs often increase their fluid intake up to four times that of their healthy neighbours, therefore the water route increases the chance of the sick animal getting the medication. They thus receive the medication quicker and could recover faster.

- Compared to feed medication there is greater flexibility in introduction and withdrawal, thus longer-term savings in medication are possible. Also statutory withdrawal periods can be satisfied easily, as cutting out the medicated water is usually simpler than providing unmedicated food on large units supplied by conveyor or pipeline.

- The veterinarian finds it easier to control the products used so there is less chance of prolonged medication causing drug resistance.

- Flexibility. There are four separate options to water medication.

  (1)   A stock solution is made up and applied direct to the feed of (say – a pen of) target pigs by watering can. Rather laborious, but precise.

  (2)   Direct application by header tanks, calculated on a mg/kg liveweight basis on the calculated water intake (100 ml per kg liveweight per day is not too far adrift). Limited to whole sections on the water line. Also laborious (header tanks can be awkward to access), and not nearly so precise. Personally, having tried it, I don't like it at all!

  (3)   Direct application to troughs. The total daily dose is calculated and put in the drinking trough through the working day. Precise, but again laborious.

  (4)   Automated equipment. Here a specific amount of a stock solution is injected into the drinking water. Dispensing automata are better these days (more rugged, less subject to damp conditions) and it certainly takes the hassle out of the other systems, and the manufacturers promote their relatively expensive equipment well.

  Good for them, but in my clients' experience, they do tend to play down the following:-

  Drip loss from drinkers is costly. Some, (e.g. lactose-based) medications promote yeasts which 'fur up' pipes and nipple drinkers.

Electric dosage pumps can malfunction in the harsh environment inside a piggery. Continuous medication (which automation encourages because it *is* so simple) can affect water palatability. Some dose proportioners require a really good head of pressure – check this out with the manufacturer before you buy.

- The water route is not necessarily limited to medication. For example, electrolytes, probiotics, flavours/sweeteners, certain types of sterilants and organic acids are other therapeutic agents which auto-dispensers can handle. But again, check with the manufacturers of both machine and drugs before you use them.

- Cost is a difficult one! On a weight for weight basis water-soluble medications are more expensive than in-feed medication. But again, as a general rule, in-feed medication lasts for 2 to 3 weeks compared to typically 3-5 days via the water, so the costs tend to be not all that dissimilar. Also, as water medication products grow in popularity so the cost difference will narrow.

- Speed of application. This always has been the big advantage of water medication over feed. With diseases of confined animals, rapid treatment is vital. I deal with this in more detail under the disadvantages of in-feed medication.

Water medication – disadvantages

In addition to those mentioned above, you should remember the following:-

- The pig's water intake is more variable than its feed intake. *This is especially true of sows* which should not be medicated via the water – but in the feed, by injection, or spot treated on the daily food allowance.

- Mixing a drug in a header tank often stirs up sediment which in some cases could 'go for' the medicant – and in any case could block drinkers.

- Canister medication (a useful option if only a pen needs treatment) can run dry, and water deprivation itself is very dangerous, especially to a sick pig.

- Again, speaking generally, experience suggests that the water route is best for enteric diseases, due to the dehydrating effect of diarrhoea leading to thirst.

  Some acute respiratory diseases, however, cause lethargy with pigs unwilling to rise to eat or drink; water consumption can reduce by 90%

and food to 100% of normal. Injections are then essential. Consult your veterinarian before hopefully medicating water they will hardly drink.

Feed medication – advantages

• Many producers are not equipped to medicate via the water.

• Undoubted convenience, and much less hassle, if ordered from a feed manufacturer.

• Many feed compounders, anxious to retain regular repeat feed business often don't pass on part of – or all of – the interruption cost of a special mix, only the wholesale drug cost, spreading the true cost of one medicated feed across all their feed orders. (When costing diets and maintaining margins during my time in a feed compounder's mill office, I had to do the calculations myself, so I know the situation only too well!)

• Again speaking generally, low-level preventive medication is less costly and easy to give in the feed. Also, while pig farmers waste 6% to 7% of food, some 65-80% of water goes to waste; expensive if it contains even a low-level drug!

Feed medication – disadvantages

• Sick pigs may eat no medicated feed, or insufficient amounts of it, defeating its purpose.

• Unless a stock of pre-medicated feed or premix is available, treatment can be delayed 12-36 hours. This is a major drawback if it occurs.

• Healthy pigs in the pen (or a whole house if on a conveyor distribution) eat more of the medicated food; thus much is wasted, going down the wrong throats.

• It is the feed mixer's task to mix and proportion the medication. Will he always get it right?

• Some compounders find special mixes a nuisance, paradoxically in a modern automatic feed plant which is programmed to run on long production runs. Other mills do specialise but may not be near you.

• Bulk bins cause giant headaches! There may not be a bin spare for a special delivery and "eating through" a bin of non-medicated feed gives great problems. For example, I have seen, when the producer is hard-pressed for bulk bin capacity, medicated food added on top of the standard ration, causing it to mix unevenly with the medicated food and the pigs receive medication at a lower rate than calculated. This both lowers the required dose to the pig and increases the risk of drug resistance by the pathogen.

• Even if there is a spare bulk bin available, problems with withdrawal regulations are more likely with two feeds being needed per house.

• Medication residues can remain in dry feed/pellet-conveyor systems and there is a risk of cross contamination to other livestock for which the medication is contra-indicated (dangerous or unsuitable).

Injections

Expensive, laborious, time consuming and can be hazardous, but essential in certain critical circumstances when individuals are very ill or will neither eat nor drink to any degree. Injection rules and techniques – which are many and important – are not covered here.

Some water medication tips

Finally, to revert to water medication, Mark White, one of our best British pig veterinarians, recommends Table 2 as a guide to getting drug proportions right via the water supply.

Table 2. CALCULATION PRINCIPLES FOR WATER MEDICATION

Volume of water consumed daily (litres) = (average weight of the animals x number of animals) x daily water consumption
Quantity of soluble product to be used = (average weight of the animals x number of animals) x specified dosage (g/kg or mg/kg)
The daily water consumption of animals is on average about 8% of the liveweight of that animal. However, this average varies with the age of the animals as well as with the season or temperature of the environment. One can accurately measure the water consumption via a water meter.

Source: White (2000)

There are guidelines for ensuring a good mix in any stock solution of medicant to be diluted/metered into a water supply. Make sure the manufacturers (Dosatron is a good one) are consulted first, and use your veterinarian as a back-up to ensure you are doing it properly.

1.   Use tepid water between 20°C and 30°C.

2.   Know the pH of the water. Some drugs like it acid and others react to 'hard' water, dissolving much more slowly.

3.   Add the drug to the water, not vice versa.

4.   Let the stock solution rest for 30 minutes. Stir well.

5.   Only prepare one day's needs at a time.

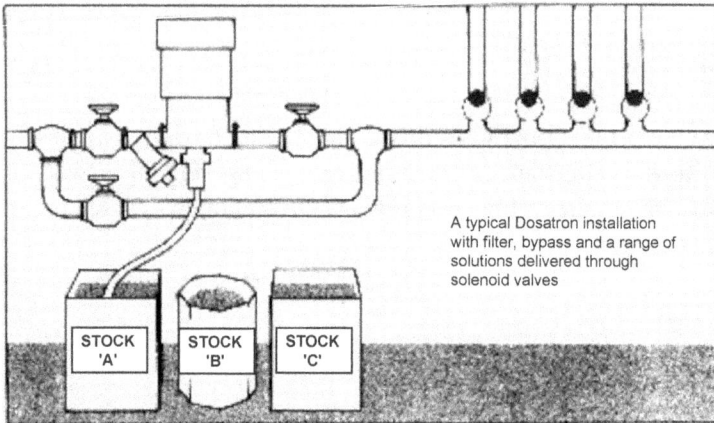

A typical Dosatron installation with filter, bypass and a range of solutions delivered through solenoid valves

A Dosatron Injector kit plumbed-in.

Reference

Pharmacia and Upjohn, 'Which Method Do You Use?' Animal Health Business, July 1999.

BIRTHWEIGHTS

This piece on a very important subject (and one which many breeders think they can do little about) was written some while ago and on re-reading it I think it has stood the test of time very well. I have hardly altered it at all.

After reading it you will surely be convinced that you *can* improve your birthweights!

Most veterinary and pig management textbooks don't cover the important subject of good birthweights all that well, in my opinion (poor <1.1 kg; average 1.25 kg; good > 1.4 kg). There have been some really excellent pig textbooks published over the past 4 or 5 years, but look at the number of pages devoted directly to this important subject - ½, 1¼, 2, 2½ in four of the volumes with, on average, 210 pages – some with 600, no less!

Birthweights deserve more cover than this.

Quicker growth to slaughter

There is ample research evidence to show that good birthweights not only provide more and much heavier weaners, but more important – that days to slaughter reduce steadily in proportion to the rise in birthweights. My attempt to plot this from my own clients and from British and American trials since 1997 is given in *Figure 1*.

More saleable meat/tonne feed from faster growth

A good deal of my on-farm work across the past 10 years has involved getting lack-lustre birthweights better – in fact a third of my call-outs since 2000 have involved this area in one way or another.

Careful records on some 30 farms suggest that for a 0.15 kg (150g) birthweight improvement, days to 100 kg are reduced by 3 days. 3 days less food gave 6.5 kg less food required per pig. At 4.96 pigs per tonne this saved 32 kg of food for each tonne purchased or an extra 12 kg saleable lean meat (MTF) produced from each tonne of grower/finisher food. This is equivalent to buying all the grower/finishing food needed 9% cheaper.

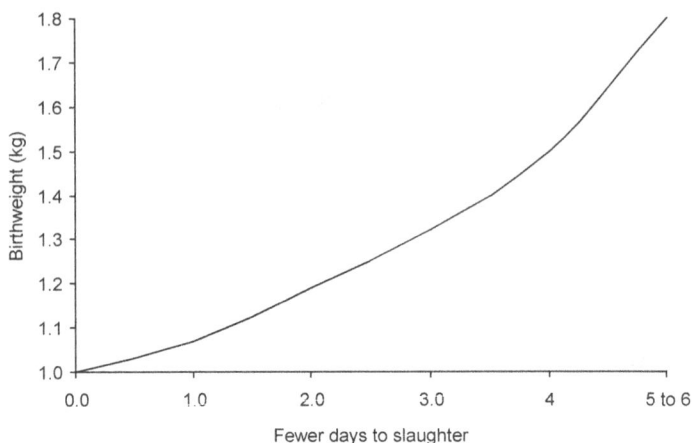

Figure 1. Projected reduction in days to slaughter (100 kg) from increased birthweights (base = 1000g).

Lower pre-weaning mortality

More piglets kept alive is a major contributor to increased income, as we shall see. Attention to the factors influencing birthweight on the 30 farms reduced pre-weaning mortality by 2.5% on average, or 9.8 born alive/litter, thus providing another 0.23 more weaners/litter. At slaughter this was down to 0.21 more pigs sold per litter. At 2.28 litters/year this is 0.48 more finished pigs per sow per year. At a 75% nominal KO this is another 36 kg lean meat sold *for each sow in the herd – some three times more profit potential than from the faster growth/ food savings advantages*.

   This is a point none of the textbooks ever make. But does this usually happen? The three times difference suggests lower mortality is indeed a major advantage, and as preweaning mortality is one of those statistics which stubbornly refuse to improve on so many farms (it has stuck globally at around 12% for a decade), the likely effect on profit of even a slight 100g rise in birthweight is invaluable in terms of more of the bigger piglets surviving.

Average birthweight is a no-no!

Up to here I've broken one of my own rules and quoted *average* birthweights. 'Beware of averages,' says the wise statistician, and he is dead right. I stand corrected!

   However, many researchers still quote birthweights (when they trouble to measure them) in 'average' terms. To do so is to miss a major potential benefit identifying and rectifying low birthweights.

It is the spread of birthweights which matter, not the litter averages at the foot of the column!

Look at Table 1, which compares recent results from two excellent British farms. They recorded average birthweights of about 1.4 and 1.5 kg. Excellent! Now I don't think a stockperson can distinguish 100g difference in this birthweight area – I certainly can't – they both look 'big enough' and there is a tendency to say 'all is well' or ' If it ain't broke don't fix it'.

But it is more broke than it looks!

Table 1. ACTUAL BIRTHWEIGHTS / MORTALITY RATIOS FROM TWO FARMS

Birthweight Category (kg)	Distribution of born-alive birthweights		Pre-weaning mortality (%)	
	Farm A	Farm B	Farm A	Farm B
Less than 0.5	0.5	1.8	80	78.2
0.5 – 0.74	2.2	1.4	62.4	63.1
0.75 – 0.99	6.2	11.8	24.7	25.2
1.00 – 1.24	16.5	20.9	13.4	13.0
1.25 – 1.49	24.1	29.1	6.6	6.2
1.50 – 1.74	27.9	24.3	3.7	3.5
1.75 – 1.99	15.1	6.4	2.5	2.6
More than 2.00	6.9	3.8	1.7	1.7
Average pigs born alive	11.7	11.1		
Average birthweights (kg)	1.482	1.366		

Despite a seemingly small difference in average birthweights (under 8%) Farm B had *the following disadvantages over Farm A*. 0.6 fewer pigs born alive; more than three times the piglets born alive under 0.5 kg, (which involves hard work to keep them alive); nearly twice the percentage of pigs under 1 kg (the ones that get crushed/chilled/scour); and more than half the percentage over 1.75 kg (the ones which get to weaning 2 kg heavier or 6 days quicker).

In birthweight terms Farm A had 0.6 more weaners/litter from their lower number of 'smalls' and larger number of 'heavies'. Thus Farm A sold about 5.4% more pigs out of the yearly sow and boar's food share of, shall we say 1.4 tonnes, to be generous. Let's also assume the farm sells, out of each sow, 22 x 100 kg live pigs year, or at 75% KO, 1650 kg saleable meat per sow per year. 5.4% more saleable meat is another 89 kg, and this divided by 1.4 tonnes is another 64 kg of meat sold from each tonne of breeding food sow used per year. This is added to the benefits of faster growth already quoted.

So what's a kg of saleable meat worth to you? In the UK a niggardly £1 ($1.82, €1.47) at the time of writing. So each extra kg of monetary income therefore reduces the notional sow food cost/tonne by that same figure. 64 kg x £1 = over £64/tonne ($116, €94/tonne). For overseas readers that is a reduction of nearly 50%. Quite an eye-opener isn't it?

This is a good example of where an economic assessment of a physical difference in performance can turn the latter on its head (see p. 43). In this case the physical difference looks insignificant. Economically it is gigantic.

Remedial action

Knowing your spread of birthweights – along with litter size, another vital influencing factor, can identify those sows which are suspect in the birthweight sector, and which can either be given special remedial treatment or culled.

In the meantime, give your birthweights an overhaul and record your spread of birthweights per litter or per batch farrowed. Too much hassle/too costly in labor? I'll cover that now.

Let's first deal with the extra costs of recording birthweights and acting on what is discovered.

Is attention to birthweights worth it?

Yes it is. Quite a proportion of my on-farm consultancy work recently has involved requests to "Try to get it better for us".

- Birthweights are important. You should know your birthweights on a litter by litter basis even though it adds a further task to a busy time. You cannot estimate birthweights by eye. 'Baby-scales' are cheap and effective. I find extra time taken is about 15 minutes per sow per year (+1.25% extra labour). This is recouped if 33% of the slaughter pigs get shipped 2 days sooner.

- Does this happen? Seems so, as Table 2 (which is quite typical on low birthweight farms) shows that careful attention to birthweights improved days to slaughter by 2.7 days for the whole herd and gave another 0.22 weaners per litter. So ***better growth rate to slaughter*** pays for the extra time needed and ***more weaner weight*** provides the impetus of greater income. In fact I calculate costs are covered some four times over.

Table 2. RESULTS AFTER BIRTHWEIGHT AUDITS, ACTION BEING TAKEN ON SEVERAL OF THE AREAS RESPONSIBLE

	Before			*After (+ 14 months)*		
	% Under			*% Under*		
	1.1 kg	*1 - 1.3 kg*	*1.3 kg +*	*1.1 kg*	*1 - 1.3 kg*	*1.3 kg +*
Farm 1						
(Birth - 88 kg)	13%	45%	42%	9%	28%	63%
Av. days to slaughter	156	151	142	157	151	141
				Av. days saved to slaughter 2.7, whole herd		
Preweaning mortality	11.7% on 9.7 b/a = 1.135 pigs lost			9.2% on 9.9 b/a = 0.911 pigs lost = +0.224 pigs weaned/ litter		

Source: Clients' Records

What to do to prepare heavier birthweights

1. **Record birthweights of all born-alives**. Some people also weigh 'healthy' born deads, i.e. not mummies, though I don't as I'm more interested in what caused the preponderance of small new-borns which lived past the trauma of birth – 'small viables'. Generally the threshold is 800-850 g.

2. **Distinguish between small viables** (as above) *and the larger* (around 1000-1100 g) *but non-viable piglets*. These are the ones which in contrast to small-viables when dried off/warmed up have little suckling reflex. Stomach tubing may – indeed will – save some of these, but the decision to take this extra trouble must rest on the individual farm and is a separate subject I cover on page 159. It is getting the *small viable* sector of the birthweight problem up to speed (i.e. improve their birthweights) which matters economically. Then the *larger non-viables* will also benefit.

Suggested targets

Small viables (and non-viables) plus larger non-viables often make up 85% of your preweaning mortality, so you can see how important it is to identify where these are, and those sows throwing them, or which farrowing conditions may be responsible. Over a spread of 10 litters at 10.5 to 11 b/alives you should have not more than 10 small viables and 5 larger non-viables of those born alive. Above this, you probably have a birthweight problem.

3.   **See if you can detect a common tendency towards small-viables in a sow, a sow line, or a boar line**. Subject these sows to a rigid examination as below:

What probably causes too many small viables – a checklist

There are many factors involved. No one seems to be sure as to which is most important or indeed, how many of the suspect causes actually work physiologically or nutritionally. This said, I have obtained success from the following: -

1.   **Persuading the producer to think about his actions well ahead of the actual birthweight record in front of him**. What he does (or hasn't done) do in the preceding parity has already influenced the result. Continuously think ahead; get into that type of mind-set.

2.   **For example:** Never let a sow nose-dive in flesh or body-fat condition in lactation.

3.   **For example:** Reduce anxiety-stress post service.

4.   **For example:** Manage sows to achieve synchronous follicle release (Synchrony = events happening at the same time).

      Theory: If the follicles are released over too long a period of time then the first to colonise the womb service at implantation secure too great a 'growth platelet' for themselves at the expense of later arrivals.

      Answer: Do a post-service (i.e. across the 24-28 day implantation period) stress check with your vet. Many factors associated with the four types of stress – fear, anxiety, actual pain and discomfort – seem to be involved.

5.   **Don't wean too soon**. Depends on the genetic strain but I find over 22-24 days helps a lot when other suggestions draw blank. Weaning too soon in some sows means that some of the womb surface (endometrium) hasn't become receptive to implantation, called regeneration, so 'clumping' of attached embryos occurs with a resultant drop in both potential litter size and a rise in Small Viables. Possibly some genetic strains can cope, others not? They may regenerate faster than others, or are less vulnerable to stress delaying the process?

What to do to get heavier birthweights

Nutrition

Don't *underfeed* lean-gene sows in mid-pregnancy. Is this practice growing? Consult the latest textbooks (like Close & Cole[1]) for guidelines on daily intakes.

---

[1] Close & Cole "Nutrition of Sows & Boars" Nottingham University Press (2000), the best nutrition textbook for breeding stock to date.

In practice, despite what the textbooks say (!), *for those sows which look as if they need it (and only those),* feed extra energy before farrowing. Then ask yourself why *did* they look peaky at that time, and get it rectified.

Follow the textbook guidelines for feeding daily intakes during the implantation period. (My clients with birthweight trouble showed a third of them were outside the advised range).

Don't feed too much whey to sows (which is a cheap food). I live in a dairying area, and it is tempting to do so.

Management

- *Overcrowding* is a possible cause. Remember, shape of pen is also important (to provide fleeing space) is as the total sleeping/exercise/feeding areas the textbooks correctly advise.

- For those of you still with sows in gestation stalls (I write from Britain!) many sows have *outgrown the stall size* of yesteryear and thus are stressed.

- Keep a close eye on all these factors on sows with large litters. These can increase your Small Viables inside an average birthweight basis, one of the disadvantages of relying solely on *average birthweights*.

- *Check on your prostaglandin technique*. Farrowing even two days too early might reduce birthweight by up to 180g and raise Small Viables by up to two a litter, so my experience suggests.

- *Serving gilts too soon*. Many European breeders don't need telling not to serve gilts before 130 kg, at the third estrus, or grow them too fast to 130 kg. But producers further afield often fail on all these counts – and my records show they often have 1st and 2nd parity birthweight problems due to uneven litter weights. If I can persuade them to slow up a bit (difficult!) then birthweights recover promptly.

- *Any disease* that can cross the placenta, e.g. PRRS and parvovirus, throws smaller pigs of those that live. Check up with your veterinarian.

- *Poor nutrition* during the first half of pregnancy causes poor placental growth, reducing birthweights.

The future

Since the foregoing was written, I have talked to several reproductive physiologists and geneticists on the subject. Perhaps it is worthwhile

to mention their latest view on the future of birthweights, which (in my opinion) has been more comprehensively covered from the dietary aspects in the nutritional textbooks I possess, than in their own field – at least at the time of writing.

HIGHER LITTER SIZE – FUTURE BIRTHWEIGHT PROBLEMS AHEAD?

We are already talking about 30 pigs reared to weaning per sow per year. I know of three producers who have tipped this scale – albeit for a short while of a year. I wrote congratulating them and got some figures in return. This meant litters of 13s, with 16s occurring periodically. Visiting two of them, the standard of farrowing house stockmanship was very high.

There seems to be a large litter effect on birthweights. One recent French trial, comparing 11 piglets or less with 16 piglets or more, suggested each piglet born within this scale meant a weight decrease of 35 grams, and piglets less than 1 kg at birth rose from 7% to 23%. Stillbirths in this group rose to 11% and 24 hour mortalities were 17%, but paradoxically with only 3% losses in the >1 kg neonates.

We have to be very careful of new, superimposed losses if we manage to move beyond 25 piglets/sow/year...

Sure, 30 pigs is a wonderful target, but it *will not be achieved* without superb stockmanship skills and enough time to practise them. And this success *will not be maintained* if, due to (many) more smaller pigs born, these skills and this dedication are not carried over into the early rearing period, too.

FARMING SOWS IN GROUPS

I'm surprised I ever wrote this article as there is so much known now about keeping pigs in large groups – weaners as well as gestating sows, and for that matter all breeding stock outdoors in paddocks, too.

But the feedback I got from it was considerable – and quite an effort was needed to reply to all the comments and queries which poured in over a 10-day period.

Obviously there are a lot of breeders interested in the group housing of sows.

Coming from Britain, which like the Scandinavian countries, has already banned the gestating sow stall, I tend to be surprised that virtually no modern textbook covers the skills of keeping sows in groups well, if at all. Over here **all** our dry sows are now housed in groups, not in sow stalls, and it seems strange that little is written about this undoubted future trend, brought about by the welfare lobby, like it or not.

*Some of us now have at least 20 years of experience (I hesitate to say 'mistakes', but as a troubleshooter in this area, there **have** been quite a few!) of keeping sows in groups profitably, and from the sow's viewpoint – comfortably.*

Problems?

At first the problems seem daunting. The sow can be a very heavy and aggressive animal, at least to other sows. Given her new-found 'freedom' to move around in gestation, when offered an inch she can take a yard. And she does! She is also a greedy old thing, and there was the question of achieving the correct daily feed intake for a wide range of body condition and submissiveness / nervousness / competitiveness within the group.

Then there were the problems of identification, inspection, selection and movement, so much easier achieved when gilts and pregnant sows are confined in a stall.

Still problems, but ....

To our credit, all these hurdles have, at last been overcome on the whole. I say 'on the whole' as if problems remain, then they can be rectified, and in my experience as a guy who is often called in to help, our mistakes today are usually ones of management (planning and lack of investment) or stockmanship (lack of time, rather than lack of skill or interest).

Those of us who have surmounted these hurdles and who are 'sailing clear' now admit – no, 'admit' is too negative a word, 'enthuse' is a better one, because enthuse they do once they've surmounted the difficulties – that the sows are so much happier, they do better (if looked after properly), productivity now outshines those sows we still see stalled in other countries (including mainland Europe) and their stockpeople are better pleased with the job – they stay on and you don't have to worry about finding replacements for skilled workpeople, which is a major headache today.

Lessons learned

I don't think you will find many of these points in the textbooks – they come from 12 years experience of three distinct phases

FIRST – THINKING AND PREPARATION PHASE (3 YEARS)

Having to prepare the ground from moving from a stall-bound breeding unit to a group-based one, and then financing and making the change. Possibly linking the major housing change with a move to depopulate, enforced or otherwise? A lot of deliberation, measuring, planning and consultation with others is needed, so allow *plenty of time* to decide what to do and how to do it.

Having then converted the buildings, you enter what I call ….

SECOND – 'THE DOING PHASE' (5 YEARS)

Learning to manage sows in groups so that aggression is minimized and pig flow maintained. Quite a learning curve which can take this long before the feeling of 'Sailing Clear' appears.

THIRD – THE REFINING PHASE (4 YEARS)

Refining the management to contain costs (bedding, food waste, identification, manure storage and disposal etc). Group housing will cost more than stalls, but it is possible to contain and reduce these initial costs with ongoing experience.

Here then, are some of the main lessons learned. I give ten or so of these, but within each category are 4 or 5 sub-sections of individual advice, making about 50 in all – too much for one, or even two articles.

Group housing sows: the basic factors

- Choose a docile breed.

- Never be parsimonious with space. Your group-housed unit will probably be 2½ times larger than your sow stall area.

- Bedding – with all its labour costs etc – is far preferable than solid/slatted concrete – even if well-made. Trust me! Do bedding well and you'll never look back. Go and see others who are successful and you will come back converted.

- For sow groups, older buildings, ***properly internally converted***, are perfectly satisfactory. Don't waste your money on 'built-from-new' palaces. It won't harm your cash-flow, so invest the money elsewhere.

- Farm for the timid sow or gilt in a variety of ways.

- Sows in yards need better stockmanship (observation, patience, agility) than sows in stalls. And about one-third more time to do a good job, but hour-for-hour there is no comparison – the result will almost always be much better. And you will be *so* proud of them!

- Use the modern electronic (ESF) feeders, of course (but there are several other systems as alternatives). If you do choose ESF, choose a rugged make, with especially good joints and fulcri, and get the layout right at the start. For example I strongly favor the twin-yard rotary system (into which most makes will fit), as it is so much easier to run[*] (*Figure 1*)

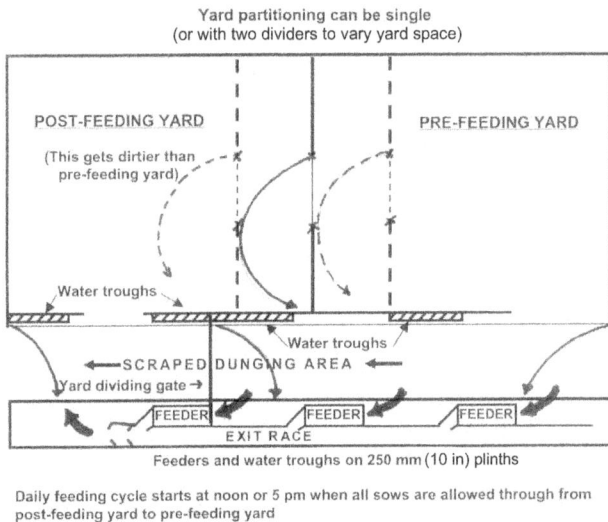

Yard partitioning can be single
(or with two dividers to vary yard space)

POST-FEEDING YARD   PRE-FEEDING YARD

(This gets dirtier than pre-feeding yard)

Water troughs

Water troughs

◄— SCRAPED DUNGING AREA ◄—

Yard dividing gate →

FEEDER   FEEDER   FEEDER

EXIT RACE

Feeders and water troughs on 250 mm (10 in) plinths

Daily feeding cycle starts at noon or 5 pm when all sows are allowed through from post-feeding yard to pre-feeding yard

Figure 1. Sketch of rotary station feeding layout for 120-150 sows

- Dynamic groups are best. Dynamic groups are where new intakes are introduced and sows removed continuously rather than the Static group

---

[*] The layout and further advice is given in my book 'Pig Production Problems' Fig 4 p 350 (Nottingham University Press, available at www.nup.com)

system where batches of sows remain together. The latter is theoretically better but difficult to maintain. The dynamic system is also quite difficult to manage, but superior when it is done skillfully and well. It is possible to run a dynamic system with size sub-groups (i.e. 'Biggies' and 'Smallies'). Generally 1000 sow herds make this idea easier.

- Gilts need special accommodation, training/mixing pens and special diets.

- **The problem of mixing sows.** There is now a great deal known about mixing sows successfully. Research this knowledge carefully (*for some of it see* p. 97). I find British and Scandinavian farmers who still get into trouble have not taken full advantage of all this hard-won experience, and it is not difficult to get things better for them when the knowledge gaps are filled.

Come to think of it, the kernel of successful sow group housing could lie in just three words …

- Space
- Straw
- Stockmanship !

Footnote: Should the sow stall be banned?

Sows in properly-designed sow stalls, well-fed and managed seem happy (not having known, or forgotten, anything else?) and productive. There is nothing wrong with the sow stall under such conditions, and its advantages of convenience, cost, and accessibility to the animal are well-known.

I got into hot water 25 years ago in my own country by saying that, in Britain, only 20% of the stall houses I visited were, I thought, up to scratch and were definitely not welfare-friendly, thus I had sympathy with the growing movement to outlaw gestating stalls and substitute a group-housed yard system for the four-fifths of our farms who were falling down on the job. I was accused of 'disloyalty'/'letting the side down'.

At the time of writing we are seeing the same arguments, especially in the Mid and North American continents.

To cut a very long argument short, supporters of the dry sow stall should come over to Europe, especially to Scandinavia, France and Britain, and look at some of our good dry sow yarded accommodation especially on straw bedding. The condition, contentment *and performance* of these sow herds is superior to the best of their stalled units I have toured, even those highlighted as the 'stars'.

Just come and see for yourselves! You will go back impressed. Sure, grouping sows is not all that easy and some of our yarded units are not so hot either, but globally the 'best average' yards are superior to the 'best average' stalls. Any day! We had to bite the bullet - forced to do so by legislation - and we are glad we did.

Think about it.

GILT POOLS

Adding in a gilt pool looks to be expensive. It is worth it?

I urge all producers to adopt a gilt pool approach to obtaining replacement females, and I need to tackle the economics of what some people consider is a necessary but expensive luxury. As far as I can ascertain for the first time carefully-recorded performance figures are put forward for you to consider. I have not yet seen these economics in any textbook.

Few pig producers today need reminding that virus diseases seem to be winning. And that every measure possible must be undertaken to redress the balance – quickly.

The gilt pool concept may not seem, at first sight, to be one of the obvious counter-measures to fight virus disease. Think about it a little, however, and you will soon see that far from being a difficult or costly thing to do – and both of these are common objections – the gilt pool is a relatively straightforward way to help make your herd immunity more robust, and which need not cost as much as you think.

What is a gilt pool ?

A reserve of properly-acclimatised replacement gilts, grown to develop leg strength and with sufficient but not excessive, backfat and muscling, correctly prepared for prompt service, and ready-and-waiting to take the place, immediately, of cull sows, whether the culling is planned into the pig flow, or is due to an emergency replacement.

Downtime is expensive

You don't need telling, of course, that culling unnecessarily late or failing to replace the cull *at once* with a properly-prepared replacement gilt raises Empty Days significantly. You know too, that for a 250 sow herd weaning 10 a litter 2.3 times a year which finds itself 20% short of properly-prepared gilts, stands to lose 200 potential weaners per year, or about 9% of output. And that this same 250 sow herd, again one fifth short of properly-prepared replacement females, automatically has its Empty Day loading increased by three days a year for all the sows in the herd. You do?

Then tell me, why is it that in the last 30 farms I have visited, 7 (23%) of these (quite good, 'top third') producers were 25% short of *any* gilt in *any* condition – let alone properly-prepared – and had to order up 'emergencies' and get them served as quickly as possible, and certainly by the second oestrus, not the normal third as preferred in Europe these days?

A planned gilt pool would have prevented all this loss of productivity.

What loss of productivity ?

First: *Downtime.* The emergency replacements weren't served for an average of 26 days – twenty-six non-productive days – on a quarter of the replacements needed. Replacement rate was 37% per year, so for every 100 sows in production at least 9 sows added 234 needless empty days, or 2.34 days per sow to the herd's annual total. At €4.50 (£3.06, $5.57) per empty day that is €10.50 (£7.14, $13)/sow added to the cost of every sow in the herd.

Second: *Returns-to-Service.* Overhasty service on such order-ups revealed 21% returns for them as against a combined normal herd average of 13%, resulting in 48 more empty days on the 100 sow average. This 14% increase cost another €2.25 (£1.53, $2.79) for every sow in the herd.

Third : *Lower Litter Size.* Further analysis of the emergency order-ups showed that their first litters were 1.1 piglets lower, and birthweights 89g lighter, sacrificing, on the litter size discrepancy alone, the profit from 12 weaners or slaughter pigs never born to the hastily-served, ill-prepared gilts. This loading would be €0.50 (0.34p, 62c) on every sow in the herd.

All these losses were measurable, but there are likely to be others.

Fourth : *Unquantifiable losses in Productivity.* Because these particular gilts were rushed into production too soon they were insufficiently acclimatised to the existing herd's pathogens. This could result in more disease, especially in the animals concerned, and a shorter productive life. Some of the 30 herds were in the tropics and we already were seeing more leg problems in the emergency order-ups, and at the last count 4% more prematurely culled i.e. after the first litter, mostly on legs with some on litter size.

Thus from the quantifiable evidence above, *not* having a gilt pool cost 7 farms at least €12.25 (£8.30, $15.20) sow each year, or for our 250 sows (the average herd size) – €3063 (£2,084, $3,792) a year.

Does a gilt pool cost that much ?

I have been involved with 10 or so conversions to the gilt pool principle in the past 5 years. The majority kept careful records of conversion and extra operating costs. Generally two group yards were supplied with wet/dry feeders, either on bedding or on concrete (bedding was by far the best). The extra housing cost, amortised over 10 years, averaged €12 (£8.16, $14.86) per gilt housed. Heavier costs came in the capital locked-up in gilt replacements purchased earlier – €37 (£25.17, $45.81) per animal. The two together raised production costs by 2½% – on a 250 sow herd €2325 (£1,582, $2,879). The fact that the cost erodes 75% of the *quantifiable* savings tends to put some breeders off the idea altogether.

So is it worth it?

Definitely! Remember there are production losses I and my 7 clients couldn't measure, and it would be difficult to see how these could fail to boost the cost/benefit gap substantially.

So I revert to my opening paragraph. If the veterinarians are right, we going to *have* to acclimatise our replacement breeding stock longer, 6-8 weeks in place of the current 3½ weeks. The nutritionist and geneticist say growing gilts can get to 100 kg in 140 days now, when 170 days is needed if legs and hormone readiness are not to suffer. Leg problems on the increase? Virus diseases besetting us? Breeding and re-breeding getting more knife-edged?

Rings a bell, doesn't it! Next I'll show you how to design and operate a gilt pool which can solve all these modern headaches. Well worth it for another €12.25 (£8.33, $15.16) per sow/year – and that is just for starters.

But before this, I want to explore two important areas a little further on which the gilt pool concept impacts.

Better long term disease status

How does the operation of a formal gilt pool improve disease? Look at Figs 1 to 4. Veterinarians tell me that the viruses plaguing us today may or may not have changed – the jury is still out on that one. "But no matter," they say "what does seem to be the case is that our pigs require longer time to achieve sufficient protective immunity." (Figs 1 and 2).

Whether this is due to current virus mutation, or the appearance of 'new' viruses coming to the fore or something else at work as yet unclear, producers know only too well that the current surge in virus attack is there. But could it be really due to our doing the induction process all wrong in the past? So that the viruses have taken advantage of it? Look at Figure 3.

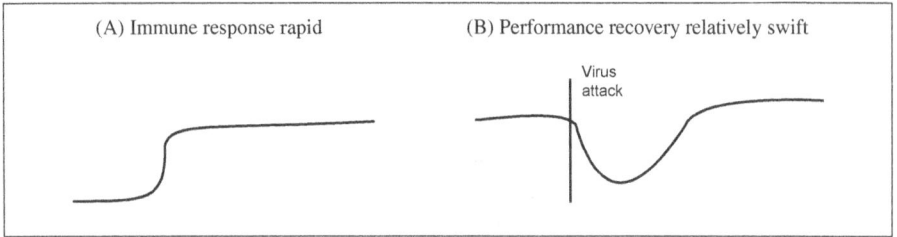

(A) Immune response rapid

(B) Performance recovery relatively swift

Virus attack

Figure 1. Previously, viral diseases had this sort of immune response to infection (A), and this sort of performance deterioration (B).

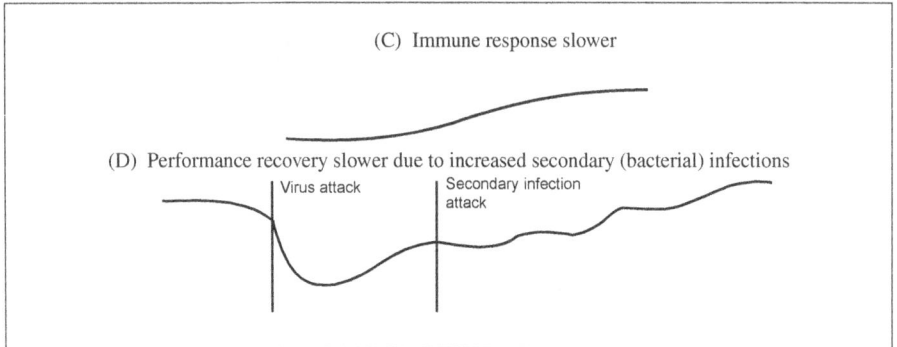

(C) Immune response slower

(D) Performance recovery slower due to increased secondary (bacterial) infections

Virus attack

Secondary infection attack

Figure 2. Now, the main reproductive reducing diseases demand longer recovery tails which may be due partly to a longer, harder climb up to full immunity (C) *and/or* being more favorable to secondary bacterial attack (P.R.R.S. and P.E.D. in particular)

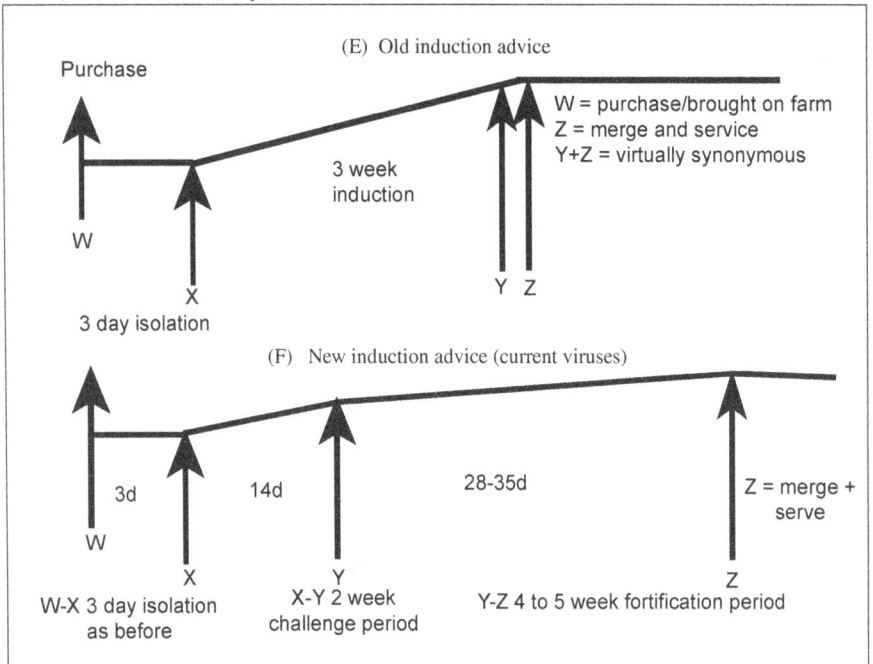

(E) Old induction advice

Purchase

W = purchase/brought on farm
Z = merge and service
Y+Z = virtually synonymous

3 week induction

W

X

3 day isolation

Y Z

(F) New induction advice (current viruses)

3d

14d

28-35d

Z = merge + serve

W

X

Y

Z

W-X 3 day isolation as before

X-Y 2 week challenge period

Y-Z 4 to 5 week fortification period

Figure 3. Infection can be materially discouraged **by longer induction periods** especially as no really effective vaccine is available for many viruses.

Old advice

What we've been doing is to follow an induction process which is too short and too rigid (Figure 3(E)). 3 days total isolation followed by 3 weeks of fence-line exposure to either cull sows/boars or weaner pigs. And/or to (in theory) boost immunity by using afterbirth or faeces – even to feeding ground-up pigs' guts which have died from an enteric infection. In other words, after the isolation period, following maybe just one protocol for too long and for too short a period of exposure-time overall. And not realising that the protocol used could now be outdated thinking in the current circumstances. (See 'Feedback and Immunity').

Latest advice

What is now advised is (after a short total isolation period as before) to split a longer acclimatisation period (6-8 weeks) into a 14 day 'challenge' period and a 28-35 day 'hardening-off' period before the new intakes are fully merged into the herd [*Figure 3 (F)*].

The 'challenge' protocol is not fixed, and depends on your current disease profile and to a certain extent from the diseases in the locality. The vet must be used to disease profile your herd and advise on the correct protocol to use - which may involve some, all or none of the acclimatisation measures suggested above. This done – and the measures could change as the farm's disease profile changes – then a hardening-off period is essential, longer for some diseases such as PRRS and never less than 4 weeks.

But what does a gilt pool have to do with this?

The gilt pool lessens the pain

Having a gilt pool helps defray some of these increased costs. First it dissuades producers (particularly outside Europe) from growing replacement females too quickly from reception to 100 kg. A modern gilt can get to 100 kg in 140 days – if fed like a grower! Result : leg problems and poor reproduction if mated too soon, i.e. "when she looks right". She may look 'right' and ready to be bred but, in human terms she is a nubile 20-year-old young woman but with the reproductive hormonal development (and the leg strength) of a 13-year-old schoolgirl! Trouble ahead in both departments! Producers, in my opinion, when buying modern European gilt genotypes, should consider the type of weight-for-age scale in Table 1. Seek advice on the exact scale from your breeding company.

A gilt pool allows you to do this, gives you time. It also allows you to buy replacement females sooner and lighter and feed them on a special gilt developer diet.

Table 1. GILTS: SUGGESTED TYPICAL WEIGHTS FOR AGE FOR MODERN HIGH
LEAN GAIN EUROPEAN BREEDS *

Gilt growth rate (lbs)	*Aim to achieve 100 kg in 170 - 180 days, gilt growth-rate at 550 g/day, rising to 600 g/day toward puberty*			
100 kg	(220)	180 days	25th or 26th week	6½ months+ old
104 kg	(229)	187 days	week 27	
108 kg	(238)	194 days	week 28	7 months old
112 kg	(245)	201 days	week 29	
116 kg	(256)	209 days	week 30	
120 kg	(265)	216 days	week 31	
125 kg	(276)	223 - 225 days	week 32	8 months old

* Consult your seedstock supplier for actual targets.

FGM (Farm Gilt Multiplication)

But there is a cheaper and possibly better way of using a gilt pool. This is to
use the extra housing needed to accommodate your own grandparent nucleus
mini-herd and breed your own replacements under the guidance of the seedstock
suppliers' genetic advice. People have been doing this quietly for years. Quietly
because the seedstock companies would rather sell gilts through their multiplier
satellites, but if pressed and at the risk of losing all your business to a competitor,
they'll sell you some good genetically-improved G.P. females, and the cross-
breeding advice to accompany them, with A.I. semen as well if you want. This
way you rear your own parent gilt replacements. The records of clients of mine
suggest this is at least €38 (£26, $47)/gilt (about 13%) cheaper than buying in
parent gilts, but as Table 2 shows, the risk of bought-in disease is dramatically
reduced from the far fewer outside replacements needed.

Table 2. NUMBERS OF REPLACEMENT BREEDING STOCK NEEDED STANDARD
COMMERCIAL HERD AND FOR GP ON-FARM MULTIPLICATION

Commercial sows	*Normal yearly gilt replacements*	*Size of multiplication herd*		*Yearly replacements*	
		GP sows	*GP boars*	*GP sows*	*GP boars*
50	17	3	1	1	1 every
75	25	5	1	2	2 years
100	33	6	1	2	1
125	42	8	1	3	1
150	50	10	1	3	1
175	58	11	1	4	1
200	67	13	1	4	1
225	75	15	1	5	1
250	83	17	1	5	1

Source : Cotswold (1988)

The €38/gilt savings is almost exactly the same as the €37/gilt extra costs of running a gilt pool discovered by my 7 clients so, with FGM, your gilt pool theoretically costs you nothing as it is part of the nucleus GP section of the farm which is paid for (across, say, 12 years amortisation) by the savings on not buying gilts from a multiplier.

Practical advice

Finally some practical tips.

Deciding on gilt pool size depends on many variables. There is not space here to describe them, but I refer you to Drs Dial & Polson's masterly step-by-step protocol covering all eventualities in International Pigletter Vol 12 N°. 2. It has never been bettered.

Gilt pools generally have to be of dynamic form (animals added and withdrawn from a group) so mixing to reduce stress is important. A suitable stress-minimal pen design for 10 gilts is given in Figure 4 (Source: MAFF (now Defra), UK).

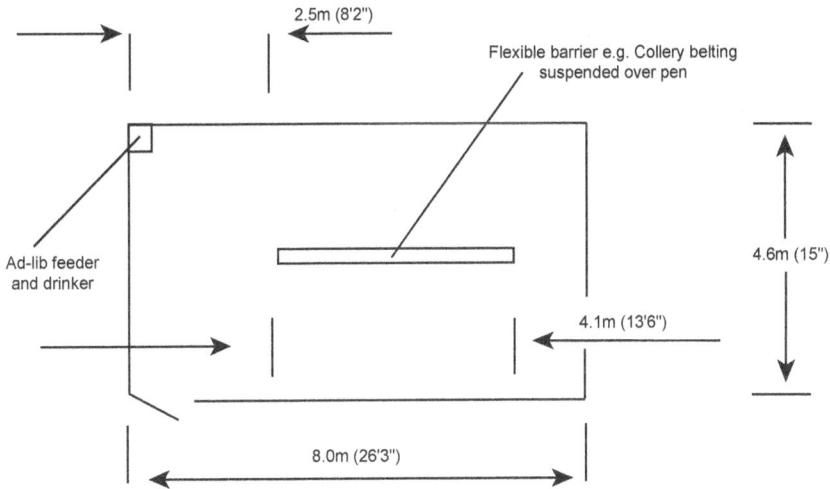

Figure 4. Suggested layout for a mixing pen suitable for 10 gilts

Introduction of sub-groups into dynamic groups

In dynamic group systems a, say, weekly sub-group should have an established hierarchy prior to being added to the main group, i.e. the use of a specialised mixing pen is recommended.

- The size of the sub-group is important and should not be less than 3 animals according to the best information available. The size of the overall group is not critical, apparently, but I'm old-fashioned enough to say six is enough – 8 at a pinch, but if the shape is right, as in Figure 4, then 10 seems to be fine, from those I've seen. But I'm a cautious old soul!

- In deep straw yards where the sub-group is added to the larger group, layout is not critical as long as there is adequate lying space, and extremes of pen shape are avoided.

- Breaking up the lying area with divisions around the perimeter – forming a specific area for each sub-group – can be beneficial particularly in non-bedded systems.

- Penning a new group in an area which has been closed off for a few days prior to the addition of the new sub-group can help in the establishment of that group's territory.

- In competitive feeding systems e.g. floor feeding, it is particularly important that newly introduced sows can compete effectively for resources, e.g. lying space, feed, water, etc. The housing/feeding system should be designed such that there is adequate space/feeding places for all stock.

- Bedding is always preferable to protect from arthritis later in life. Keep the pen shape as square as possible. Do not overstock, and provide fleeing space. Provide snug hovers over the sleeping area in cold weather.

- On hard floors a few partitions to encourage family grouping early on is useful *as long as the resting gilts can see the feeder (vital)* or feeding space if dump feeding (less good) is practised.

ON-FARM TRIALS

Among the badly-needed textbooks yet to be written is a simple introduction to statistics. Yes, I know, there are dozens on the subject in the bookshops and libraries, but I've never managed to get past any author's chapter two before I'm trailing behind in comprehension as the subject description gathers pace at the speed of a rocket off the launch pad!

So it is with 99.9% of farmers, I fear. Yet it is these same farmers who are bombarded with trial 'proof' of the benefits of some product, idea or system which the vendor's sales and marketing departments claim is better than anybody else's.

Question. But *are* they? Anyway, how do you tell? They can't *all* be best!
And how often do you read in the agricultural press the remark by a questioning pig producer ".... So we decided to set up our own test under our farm conditions ..."

Question. But was it set up properly and analysed correctly even so? Sadly, in four-fifths of the farm trials I've encountered – it hasn't been, so it was largely a complete waste of time and enthusiasm, as I shall show in a minute.

Good guidance from the experts

If you look at excellent articles by experienced research workers – eg Prof Colin Whittemore ('How Not to be Misled on New Knowledge') and Prof Duane Rees ('How to Design On-Farm Feed Trials') to name two of the half dozen I have here before me, the farmer's immediate response is to think, 'but I can't do *that*!'.

The second reaction is to say, 'so there's not much point in my doing a farm trial'. (This, by the way, would be accompanied by a fervent 'hear, hear' from any academic, as all of them have experience of the four-fifths of farm trials which indeed weren't worth the time and effort the pig producer put into them.)

Do it properly or not at all

The on-farm trial is worth doing – properly – *firstly* because conditions on individual farms are so very different in bacteriology, virology, microclimate, genetics, investment, management, pig-flow and what I call labour-flow – that no one product, idea or system will necessarily give the same result on every farm.

Secondly: there are so many options to choose from these days – take for example the topical subject of antibiotic growth promoter (AGP) alternatives. A farm trial – or a series of properly conducted ones across a period, can at least narrow down the choice on which of them it seems best to try. And yes, several trials may be needed as conditions change on any farm.

Third: the best pig producers I know always have had a portion of their farm – and of their time and energy – as an on-going test-bed. Feed, genetics, equipment and different management options. This helps motivate the staff if the trial is properly explained, and it has thrown up some clear guidelines in both the success and failure areas from which everyone learned something.

So I give below (humbly, as I'm not a statistician) what I consider the textbooks should be telling you – in the form of a basic matrix – on how to do a farm trial which will tell you something worthwhile. And remember; negative information can be as valuable as a positive result, as it will tell you what not to do in the area examined. This will stop you wasting future money and time.

An on-farm trial which is worth doing

FIRST – GETTING COMMITMENT

What will make you change?   Decide on a profit figure, be it margin over food cost, gross margin per pig sold, cost/kg live or deadweight gain. You choose – it's your decision – you know your business best. **A tip**: quite a good one (for feeds) is kg of saleable meat per tonne of feed (MTF) as this takes into account pig performance, economy of gain and return on carcase (grading), which can all be related back to several profit yardsticks. It also includes a major rule in a feed trial, in that we want to pick up what differences there are, if any, *between the feeds and nothing else.*

So... what will make you change in this respect? 1 kg of extra saleable meat/ tonne feed – not worth it! OK, 10 kg/tonne feed? Maybe, but hardly. OK then, 25 kg? That's more like it, as at 5 pigs per tonne (roughly) this is 5 kg more lean per pig, worth having, you say, for the same input.

SECOND – BEING CONFIDENT OF THE RESULT

Right! Let's now design a trial so that if there is 25 kg difference per tonne of food consumed between Food A (your present feed) and the best likely alternative, Food B, the trial will reveal it to a degree i.e. a "confidence limit" which is itself reasonable. There is no point in having a food which provides a fantastic result if the statistics suggest it will happen only once in a blue moon, or even it happens 10 times out of 20 – if the other 10 occasions give a result as equally bad as the result was good, then you've gained nothing. See what I mean?

THIRD – FINDING OUT WHAT YOU NEED TO DO

We now have to submit our needs to a statistician – most research scientists are trained to this end. You will need to approach the agricultural division of your nearest university. (Advice – best to get a trained animal researcher to do this for you, not a 'general' statistician.) Of course you'll have to pay for this, so clear the cost first by visiting them. They will say that to achieve that sort of difference, say, 15, 16 or 17 times or more out of 20, you're going to need *x* number of pens on Food A comprising *y* number of pigs per pen and similar numbers and pens on Food B.

Sometimes, at this stage, the farmer's face pales as he is told of the large numbers needed. Quite impractical for him. Don't panic!

FOUR – ADJUSTING YOUR TARGETS TO SUIT YOUR CAPABILITIES

The statistician can discuss with you how much to lower the target difference you have chosen so as to be able to reduce the number of animals and pens (which latter they call 'replicates') which you can handle. This then makes it a feasible trial to carry out from your point of view. He may lower the confidence limit a bit if this helps, but still expose the performance or cost-effective difference in the two feeds if it exists.

However, *sometimes* the two don't meet! The scientist insists on too many pigs; you won't change for what you consider to be a minor improvement cost-wise – although in fairness *any* cost benefit is worth having these days; but sometimes there are other factors of convenience or credit to be taken into account, too. Don't argue! If you cannot agree, then call it off. Not to do so wastes everybody's time.

FIVE – DOING THE TRIAL

Having agreed to do this (or not – if not, then don't go ahead, remember), all you have to do is carefully carry out the trial instructions you will be given as part of the service.

Generally, pig farmers and their staff do this conscientiously and well.

SIX – ANALYSING THE RESULTS

But it's not over yet! Having concluded the trial and got the data, this has to go back to the scientist or to a qualified statistician to be analysed. They will tell you **what you can and what you cannot say about the figures**. If you are lucky you will have achieved your pre-fixed 'worthwhile-to-change' target. Usually not, I must confess, because farmers' targets tend to be too optimistic. But all is not lost, all the same. A good analyst will be able to discuss with you the merits of various aspects of the trial as it affects your business. If the statistician cannot do this, not being a trained agriculturalist, then an experienced animal specialist can do so – the important thing is to dialogue with an independent expert **and not with the manufacturer** as they are likely to be biased. No – they don't tell lies, but it is only natural in a selling situation not to highlight the blemishes, only the plus points.

SEVEN – THINKING ABOUT THE RESULT

While I prefer a 'make-you-change' threshold to be an econometric one (i.e. taking account of cost-effectiveness) many producers just work on physical performance differences like FCR, or % first grades/P2 measurements/lean in carcase, or days to slaughter etc – usually a combination of them. That's OK, but do relate them, first, to what margin you would be likely to achieve; and second, in relation to the amount eaten and cost per tonne of the two feeds or feed additives.

Quite a performance!

I agree, it is. And farmers don't like paying for the **essential** statistically-correct 'setting-up' and 'analysis' advice anyway, but I have seen at least 20 farm trials where the time and effort expended in blissful ignorance of what is needed has been 2 or 3 times more than this cost. And all to *no* effect, as inevitably the conclusions at the end of the work were shewn to be within the laws of chance anyway. What a waste!

As to the cost of using a statistician, see my section on the subject on page 103.

So do a farm trial – but do it properly!

FARM TRIALS;
IS EMPLOYING A STATISTICIAN TOO EXPENSIVE?

I wrote this piece several years ago, and note with growing concern that no textbook which has been written since then has tackled the subject. I still see farm trials quoted which really don't stand scrutiny – at least in my opinion, for the reasons quoted on page 99.

So I am very happy to republish this particular essay as it stands, and renew the plea that all serious textbooks quoting results should describe how to do a field trial properly.

Open any pig text-book you possess – or consult one in a library, and not one of them that I've seen has a chapter on doing farm trials properly!

Considering that livestock textbooks are usually written by scientists who are the one group of people who can give sound advice on this subject, I find this a surprising omission.

I can only put it down to the authors not wanting you to even attempt comparative on-farm trials, because past history shows that 85% of pig farmers get it badly wrong and the results they are proud of are at best only a guide, and mostly are statistically meaningless. In other words the results are usually within the laws of chance and thus a waste of effort.

Chicken and egg situation ?

How can we break this impasse? I guess farmers get it wrong because scientists don't tell them how to set about doing farm trials properly in the first place. And farmers never think to ask them anyway. If both of them don't get together before the farm trial, how can farmers be expected to get it right in the end? A chicken-and-egg situation!

You must use a statistician

It all hinges on statistical accuracy. Farmers can do the donkey-work of on-farm trialling perfectly well under guidance – I know this from helping them carry out dozens of on-farm comparative trials on food, additives, systems and equipment over the past 20 years. Bless them, they all did it well.

But every one of these farmers has had the help of a statistician, generally arranged for them by myself.

All scientists have been trained in this subject, at least sufficiently up to farm-trial level as these usually involve simple two-way comparisons. Professional statisticians are employed for the more intricate three or four-way comparisons used predominantly in a research situation.

Statistical calculations are needed to **set up the trial properly**, and to analyse whatever results appear at the end of it so as to outline **what you can and cannot say about the results**.

Framework for a farm trial

Put simply, the way I approach a farm trial is like this.

1.   We want to compare, say two feeds, on econometric grounds (Econometric = measurement of cost-efficiency)

2.   First, decide on what extra income, or profit, or other financial incentive you prefer such as Margin over Feed Cost (MFC) **will make you change** from the feed you use now to the trial feed. The "will-change threshold".

3.   Employ someone trained in statistics to tell you how many pigs in each trial group are needed to reach your "will-change threshold" and how the details need to be organised to give a confidence limit (if it is there) of getting a positive result 17, or if you prefer, 18 times out of 20.

4.   Very often the producer is put off by the numbers and the time and effort needed. Don't argue! It is better either to reduce your "will-change threshold", or reduce the confidence limit to 17 times out of 20 (no lower) or abandon the trial so as not to waste your time and effort, or take a deep breath and do it anyway to meet the statistical rules demanded.

5.   Do the trial.

6.   Submit the results to a statistician who will tell you whether your "will-change threshold" has been reached and how far you are adrift if not.

7.   Decide on go or no-go on making the change.

Statisticians? Too expensive !

How often have I heard this. As a result, some quite promising new products or feeds which never got trialled because the farmers objected either to the cost of the statistician or the 'botheration factors' he laid down, I have recorded the actual cost of statistical work on 11 farm trials I have been involved with.

6 of these trials gave positive results, and 5 negative ones, negative being that the product trialled didn't reach the farmers "will-change threshold". By the way, before starting the work, 7 of the 11 farmers had to revise this "will-change threshold" downward, usually by 10-15%. Producers seem to be rather too optimistic, preferring at first to increase gross margin by 25% rather than by 10%! These days, it doesn't happen in real life.

Table 1 is interesting if only because I have never seen data like this before.

Table 1 HOW MUCH DOES A STATISTICIAN COST?

The result of a farm trial is statistically positive, negative or inconclusive. Here are 11 farm feed trials where six were positive (change made) and 5 negative or inconclusive (no change was made).

'Positive' result. Cost of statistician in relation to 12 month's increased gross margin (4 cases) or increased income (2 cases)		'Negative' result Cost of statistician in relation to current annual margin over food costs/pig	
Farm	*%*	*Farm*	*%*
1	36%	1	1.33%
2	1.6%	2	0.81%
3	10.2%	3	0.56%
4	0.7%	4	0.89%
5	4.1%	5	0.5%
6	3.0%		
	Average 9.26%		Average 0.82%

Source: Clients' records (UK) 1988-2003

Conclusion

While 11 farm trials is a small sample and in this case the cost of employing a statistician to set up and analyse one feed trial may seem high, these results suggest that if the result is statistically favourable his cost could still be under 10% of the first year's benefit gained, and even if negative or inconclusive you may only have 'wasted' under 1% of your current margin over food cost by using him.

What does this mean ?

Let's assume you do 3 on-farm trials a year, two of which are negative, and one positive, the positive one suggesting that you will improve your margin over food cost (MFC) by 10%, i.e. gross margin improved by 6%, which in my experience is attainable.

By employing a statistician to do things properly for you you will have lost 2% of your MFC in two trials but gained 10% in the third, a total benefit of 8%, or a 5:1 REO.

In nett margin over feed cost terms the value of using a statistician, even if two-thirds of the trials in this example proved negative, this would raise the MFC by 6.87%. That's good, guys!

I submit that using a statistician is a good bet financially as well as having that peace of mind from not having wasted your time and securing results from which you can make important decisions.

My advice is brutal and blunt – do not do a farm trial without a statistician to design it and interpret the results. If you don't you could be wasting everyone's time – especially your own ! And it doesn't seem to cost as much as you think.

YIELD
(CARCASE DRESSING PERCENT; KILLING OUT PERCENT)

Yes, textbooks mention yield of edible meat, but maybe not all that well!

Considering it is so important to us I feel they do not cover it sufficiently – only one of the major textbooks I've read does so, but even this one not really comprehensively **in practical terms** – the theory it describes is magisterial however!

Yield differences of 3%

I find, on the farms I visit, differences of up to 3% but 1.5 to 2% is commonplace. On a 110 kg live finisher pig 3% is 3.3 kg less edible meat sold – or 6.6 tonnes less meat per 100 sows progeny annually! Even on a lighter 95 kg live pig it is 5.7 tonnes! That is a *huge* amount which slips away unsecured!

With profits so tight these days, we just cannot afford even half of that shortfall, so I'd like to draw some of the loose threads together, which other textbooks seem to have left dangling.

Case histories

Quite a few years ago I was asked to look into low KO% on three farms all at or around the 72.5% to 73.5% level. By attending to only 5 of the items listed below, over the next 24 months we increased yield of edible meat per farm, per progeny of 100 sows, by 2.97 tonnes, 3.60 tonnes and 2.04 tonnes respectively! All this from increasing yield per carcase by an average of only 1.6%.

The factors primarily attended to in all three cases are of interest because they certainly worked, and are still continuing to do so, the producers tell me.

1.   Closer liaison with the abattoir; regular visits made – yes to check on their handling, lairage time, watering and grading procedures. This was in the UK and they didn't mind one bit; indeed they welcomed the 'interference', saying it wasn't all that common, many farmers just didn't bother to take an interest.

2.   Arrangements were then made for twice-weekly shipping so that....

3.   ...Closer (i.e. heavier) weights to the maximum permitted on the contract were achieved. On one farm this was as much as 4.3 kg! Weighing equipment was partly updated on 2 farms to make the job easier.

Weighing wastage

I know, twice-weekly shipping is a pain and it is certainly more expensive in labour and time – mathematically up to half as much again. And, as I show below, there is an alternative which my smarter clients have switched to after they found a period of double-shipping tended to upset the work-flow. 'Weighing wastage' I call it. Let's look at this a little more closely.

My table at the end of this article shows the value of getting closer to the contract maximum weight ceiling. Heavier pigs yield more saleable meat, and while it cost 3.7% more to get the extra, the return on the heavier carcase (on this particular contract) was 6.7%. In my experience it is usually a 2:1 ratio similar to this example.

Double shipping (if you are 1 kg/carcase adrift or more) often gives a higher return ratio (2.5:1) – but it does cause mayhem with the workload!

A better way?

Yes, there could be. This is to go back to shipping weekly as before, sail a little closer to the wind with the weighing, and providing the overweights are not more than 3 to 4 kg (live) and that they don't overtop the demand from a friendly pork butcher (see below) you are penalised nothing like to the same extent should you send these overweights to the contract purchaser rather than this 'niche' outlet. In fact there is usually no penalty.

The pork butcher (or butchers) are glad to get them because, if you produce succulent meaty carcases, they aren't going to bother about a 4 kg or so heavier carcase, which wouldn't fit the processor's automated processing lines, thus you won't suffer their overweight penalty.

Sure, it can be a limited outlet, but I've known localities where it hasn't yet been explored – Farmers' Markets are a case in question. So check the situation out at the various local towns, where you could secure several, not just one, such outlet.

Back now to how we improved yield on the three farms.

4.   Boar genetics were improved into meatier, higher KO% potential lines, but as well and just as important, the diets were adjusted to higher nutrient-dense, lower-fibre feeds. This itself seemed to improve the yield of edible meat per tonne of food (MTF) by 8 kg (or 1.47 kg per pig).

5.   Split-sex feeding was commenced (entires and gilts). This gave another 12 kg of saleable meat per tonne of food used.

6.   During this work we also looked at the effect of scouring (mostly post-weaning) on KO%. Pigs which scoured early in the growth period but recovered, had a lower yield of edible meat (i.e. lower KO%) compared to those which didn't scour, even if they grew well subsequently.

Factors affecting yield

Here is my list of the practical factors involved. The textbooks never mention them *all* and maybe I've missed something too, but I don't think so.

- Close liaison with abattoir
- Known slaughter time
- Ship close to maximum permitted weight
- Prompt killing
- Short journey time, especially in 'white' breeds
- Lairage water available
- Shipping twice weekly
- Last feed 12 hours before killing
- Meat-type terminal sires
- Attention to the prevention of 'nutritional' scours post-weaning?
- Blocky, compact genotypes
- Add special low heat-increment diets in hot weather?
- Correct feed matching to such genetics
- Differential sex feeding
- Use mineral proteinates (Bioplexes)?
- Check, check, check, your farm weighscales

And not castrating?   No I won't go into that one as these articles in Pig Progress are read around the world, and boar meat is really a matter for us odd British .... !

WEIGHING WASTAGE: THE ADVANTAGE OF TAKING PIGS TO HEAVIER WEIGHTS COMPARED TO THE ADDITIONAL COSTS Conversion rates: £1= $1.82, €1.47

Maximum Sale Liveweight 103 kg. One farm, two sections fed and managed identically

Section	Av kiveweight (kg)	Killing out %	Yield of saleable pork	Value (€)	Extra costs* (€/pig)
			(kg/pig)		
A	101.3	79	80.03	84.35	3.12
B	97.6	77	75.00	79.05	—

* Food, labour, power, buildings

Result: Sale value of the heavier pigs overrode extra costs by €2.18 (£1.48, $2.70)/ pig of saleable carcass meat
The advantage of shipping x 2 week compared to x 1 week
Av slaughter arrival weight 102.4 kg *v* 101.3 kg
+ €1.1 (75p, $1.36)/pig income *v* extra cost of shipping €0.40 (£0.27, $0.50)/pig
Gadd (2004)

PIG TECHNICIAN'S SECTION
Ideas for the stockperson of tomorrow

SOME THOUGHTS WRITTEN FOR THE (USUALLY) OVERWORKED PIG TECHNICIAN - A NEW NAME FOR THE STOCKPERSON

As an animal lover I have always been fascinated by stockmanship. Apart from one excellent book and one manual both on pig stockmanship, all the rest of the textbooks I possess pay lip service to 'good stockmanship being vital', but never support it with enough detail. Maybe the authors haven't been that close to pigs for long enough.

My long life among pigs – and especially among those hard-worked men and women who care for them – firstly involved doing every single job myself (sometimes well, sometimes, I admit, not so well!). I've always liked to get 'my hands dirty', which is an unfortunate phrase these days, and regard manual work with pigs a relaxation from the intense mental effort that being a farm adviser often entails.

So all my life I've learnt from stockpeople, listened to their gripes, worked alongside them, taught and trained them, motivated them, monitored, recruited them – and had to sack a few – not the most pleasant job in the world.

Good pig stockmanship is a very satisfying job and a career in itself – providing you have a good manager or owner!

ARE PIG TECHNICIANS[1] BORN OR MADE?

Made! But I've met some who should never be responsible for stock because they will never become stockpeople. You have to like animals for a start. Then you have to create an atmosphere in the piggeries of security, consistency and trust. This means making to effort to *care* for them – not just like them. Looking after animals well takes a lot of self-sacrifice.

Modern research has proved that quiet, confident stockmanship gives better production and performance from contented stock. Why?

* The pig is under less stress with lower levels of those hormones which depress performance.

* A close relationship with a pig performing naturally allows the stockperson to notice changes and take prompt corrective action. Scared, unhappy pigs tend to act the same way; in terms of productivity – negatively.

---

[1] Notice the deliberate use of 'technician' for 'stockperson'. We need to give stockpersons some more status in the community.

•   Consistency allows the animal to settle into a routine. Pigs are creatures of habit and stability encourages them to cope with varying degrees of an unnatural environment. Personally, I don't think pigs are all that intelligent, but they are intensely *aware* animals – they know what is going on around them and respond to routine/familiar things.

Concentrate on these:

1.   **Sight**   Recognizing changes in appearance of the pigs and their attitudes.

  Problems: Familiarity breeds casualness. You look but don't see. Being observant is a tremendous attribute for a pig technician to possess. *But good observation takes time.* Time is a stockmanship cornerstone which when curtailed, brings the whole edifice down.

2.   **Smell**   Recognizing changes in the smell of the pig and its housing.

  Problems: Research proves that stockpeople get used to smells (gas levels), called habituation. Certain products and 'elbow-grease' keep down performance-sapping masking smells (ammonia and hydrogen sulphide).

3.   **Hearing**   Recognizing the sounds pigs make, especially at rest.

  Problems: A surprisingly weak point. Pigs talk to you all the time, vocally and visually! A good pig technician listens, understands, acts.

4.   **Touch**   Recognizing the feel of a pig.

  Problems: Few people condition-score sows *properly* (causing it to get a bad name), disturb stools with their feet, palpate udders, tap for congested lungs, feel for hard stomachs, etc. Get your vet to show you – if he is a pig vet, that is!

  Yes I know, now we have hand-held fat and lean measuring devices, the old method of condition scoring sows is less accurate, more subjective, probably has had its day in the gestation period. Yet skilled condition scoring *down through lactation* can pick up the 'nose dive' in condition quicker than any machine, or your own eye. Know the feel of those aitch-bones beginning to come through, signalling fast remedial action.

5.   **Talking**   Pigs like to be talked to; they perform better. Unproved, but true!

  Problems: People think it is effeminate, that their colleagues think them crazy or 'soft'. Ignore them. Do it! The familiarity thing again. Passes the time, too!

6.   **Feeling**   Sensing draughts, cold spots.

  Problems: Physical work masks the ability to *sense* discomfort in the pigs, thus visual signs become even more important.

7.   **Checking**   You must install measuring devices and monitor them. Check self-feeders, temperature, fan speeds, stocking density, controllers, drinkers, disinfection thoroughness, feed allowances, etc.

Problems: Owners and stockpeople don't realise these checks are the back-up to poor or hurried stockmanship errors. A good technician checks himself, his routine, his equipment, as well as his pigs.

8. **Presence**   Now that more of us are keeping pigs in larger groups, getting in amongst them is becoming more important (see page 131).

    Problems: Not close-checking strawed sows in group yards every day. Not having someone present at farrowings.

9. **Thinking**   A good technician plans his routine and subconsciously builds in a contingency period – at least half an hour a day. (One hour for sows kept outside) to allow for things which interrupt workflow. They always do and always will. Workflow affects pig flow, affects performance, affects profits, affects job satisfaction and even personal relationships. I've had many a heart-to-heart with a technician who wants out. One of commonest *real* reasons for a stockperson quitting starts from an unsatisfactory, jarring workflow – you can often see the knock-on effect all the way through the previous list. One of the basic reasons for quitting starts from lack of time – and the more conscientious the person, the more it grates.

Self assessment

The table below lists the common attributes of pig technicians. Try this little exercise: assuming the centre line is neither one way or the other, just put one mark near or far to the left or right of it according to your own opinion of yourself. Think carefully about the opposites. Be honest. Just one mark per line.

Self-Assured	. . . .	. . . .	Apprehensive
Practical	. . . .	. . . .	Imaginative
Conservative	. . . .	. . . .	Experimenting
Reserved	. . . .	. . . .	Outgoing
Forthright	. . . .	. . . .	Devious
Serious	. . . .	. . . .	Happy Go Lucky
Tough Minded	. . . .	. . . .	Sensitive
Timid	. . . .	. . . .	Venturesome
Submissive	. . . .	. . . .	Dominant
Affected by feelings	. . . .	. . . .	Emotionally Stable
Expedient	. . . .	. . . .	Conscientious
Uncontrolled	. . . .	. . . .	Controlled
Trusting	. . . .	. . . .	Suspicious
Group Dependent	. . . .	. . . .	Self Sufficient

Some easy to answer – some difficult! Now turn to page 117. To see how you match up to the ideal stockperson.

The team spirit \'esprit de corps'

Pig farmers in my own country (Britain) have a lot to learn here compared to the larger German, North American and Japanese farms. British farmers would argue that, on average, individual stockmanship is better in Britain than in any of these countries, which is true, but they miss the point of those nations who are naturally more group-oriented. This is that team-wise, they take a great collective pride in the unit. As a result the whole farm is neater, tidier and better-organised; it is *good organisation* on the larger pig farm which is a vital prerequisite of *good stockmanship*. The larger pig farm in the future needs both. The first-class manager can improve both the organisation and the stockmanship, but of the two his main role lies in organisational skills. The key task where stockmanship is concerned is for the manager to recruit good section heads and get the labour load right, then the general pig technicians will improve by leaps and bounds, as it is a self-fuelling process.

CASE HISTORY

Measuring the effect of superior stockmanship – a rare example from North America

For an overall view of how important good stockmanship is, I describe a rare occasion where an order to a major seedstock house for boars and gilts was split (erroneously) between 2 clients of mine, one in the top third productive category with excellent pig technicians and one well down the ladder in this respect. So parts of both farms suddenly had the same genes, the same food and the same buildings in the same locality but the management and stockmanship were very different.

I kept careful records of the performance of the respective sows and their progeny from both farms involved in the unfortunate supply mix-up (*see Table below*).

Mostly Female-Related Performance	*Farm A* (*Good Stockmanship*)	*Farm B* (*Poor Stockmanship*)
Numbers weaned – 3 parities	10.64	9.06
4-week weaner weight per ton of feed used	340 lbs (154 kg)	240 lbs (109 kg) *(-41%)*

Mostly Boar-Related Performance	*Farm A* (*Good Stockmanship*)	*Farm B* (*Poor Stockmanship*)
FCR (20-196 lbs) (9-89 kg)	2.54	2.75
Saleable meat per ton feed	624 lbs (283 kg)	564 lbs (256 kg) *(-10.5%)*
Margin over food cost (same deadweight price / lb)	$343.50 (£188, €277)	$295.50 (£162, €238) (-16.2%)

What were the differences in stockmanship management?

1. (A) spent more time on breeding – 2.21 more man-hours/year/sow and gilt; about 12% more time in this area.

2. (A) spent more time at farrowing as they used prostaglandins, while (B)'s men could only attend 37% of the farrowings during working hours. As a result (A) had mortalities to weaning of only 7% (0.83/litter) while (B) had 12.2% (1.27/litter).

3. Scour was jumped on far quicker by (A)'s men, though the incidence on both farms was similar.

4. (B) rarely checked or altered feed scale; the wet:dry feeders were poorly adjusted, and the starter grower hoppers not kept "fresh".

The bottom line is that while (A)'s overall labour cost was some 20p ($0.36, €0.29)/pig) higher per finished pig than (B)'s, his gross margin on these particular "pigs in parallel" was over $4.20 (£2.30, €3.40) per pig higher, or more than a x10 payback!

Good pig technicians matter very much to both performance and profitability. *People* make – or break – businesses.

The ideal stockperson? Answers to quiz on page 115.

Dr Seabrook, an animal behaviourist, suggests a good animal attendant is like this:

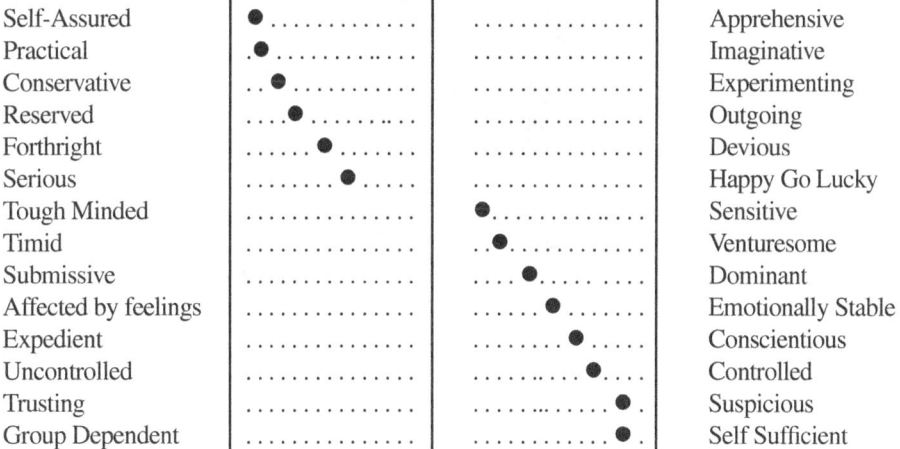

Self-Assured	●	. . . .	Apprehensive
Practical	. ●	. . . .	Imaginative
Conservative	. . ●	. . . .	Experimenting
Reserved	. . . . ●	. . . .	Outgoing
Forthright	. . . . . . ●	. . . .	Devious
Serious	. . . . ●	. . . .	Happy Go Lucky
Tough Minded	. . . .	●	Sensitive
Timid	. . . .	. . ●	Venturesome
Submissive	. . . .	. . . . ●	Dominant
Affected by feelings	. . . .	. . . . . . ●	Emotionally Stable
Expedient	. . . .	. . . . ●	Conscientious
Uncontrolled	. . . .	. . . . ●	Controlled
Trusting	. . . .	. . . . ● .	Suspicious
Group Dependent	. . . .	. . . . ● .	Self Sufficient

How close was your slope left to right? Now get your wife/husband (or even your boss) to do it, giving their opinion of you! It can vary from being a rewarding to an excruciating experience!

Several people queried the absence of 'observation' as a desirable faculty, especially as I believe it is such an important attribute of a good stockperson. But we are all extremely bad judges of our own competence in this area, as any well-trained police officer will tell you, so it has been omitted. Anyway, what is the opposite of observant? My 'Reverse Dictionary' is silent on the subject!

"Pig technicians are made, not born". Well, maybe there are exceptions!

SOWS PER MAN? ... OR MAN HOURS PER SOW?

A piece written for the pig technician. And I hope their employers read it too! The basics were written 8 years ago – I've brought it right up-to-date with Table 1 compiled this year.

People often ask me "How many men should look after 100 – 200 – 400 – 600 sows etc, farrow-to-finish?" Farmers look at it like this because they are thinking about spreading their wage bill across as many sows as possible. With respect, I think this is wrong! The pig technicians I know think in terms of spending time with the pigs, and often complain to me that they have to do too many tasks which take them away from stockmanship – like repairs, records, moving muck, mixing and humping food and cleaning down. None of these are inessential tasks, but they need lifting to some extent off their backs so they can spend more quality time with the pigs. I agree, *it is man-hours per sow that matters* not sows per man.

Look at it this way. My farm records alerted me in the mid-eighties that as sow herds got somewhat bigger, the hours spent with the pigs got less, yet the labour cost rose[1]. This stimulated me to record hours worked (as taken off wages sheets) against sow productivity (from the farm records) on about 150 farms I visited in the period from then until the mid 90s.

Very interesting!

This has resulted in the graph below taken from a wide variety of farms across the world (12 of them from very large farms in the USA). To me it suggests that there *is* a relationship between productivity and man-hours per sow per year. In other words technicians on the best farms spend the most time with the sows.

---

[1] Now that economy of scale has clicked in on the much larger pig farms we see at the time of writing, labour costs per sow are coming down again. The problem the massive units have is that insufficient time is spent with the pigs – back to the stockperson's gripe of 15-20 years ago. That hasn't changed even if farm size has.

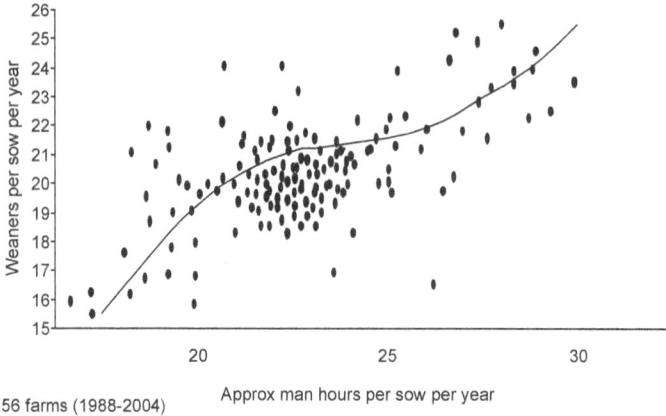

n = 156 farms (1988-2004)   Approx man hours per sow per year

Note Productivity < 18 weaners/sow/year predominantly Far Eastern & smaller U.S. farms
Productivity > 19 predominantly U.K. top third farms.

Figure 1.

How do you rate?

Pig technicians, maybe it is an idea to plot your own number of hours worked per year and divide that figure by the number of sows you look after. Then set this against your productivity record in terms of weaners achieved per sow per year.

If you find your farm is down near the bottom left-hand corner of the graph then you should question your labour allocation. Maybe you are doing too much donkey work? Spending too much off the pig unit on other farm tasks? Not enough automation to take the heavy and dirty, time-consuming jobs off you? Too much tailchasing on repairs?

Any business makes or breaks itself on its use of people. Maybe too, you and the boss need to look at how you as the stockperson spend your time? And to aim for more time being spent with the pigs.

Think about it!

APPORTIONING THE HOURS WORKED PER SOW

These results encouraged me to go on to record how the man-hours were split between tasks on another 50 or so clients visited in the past 10 years (1994-2004). Table 1 expresses them in the interesting form of man-hours per sow. I've seen similar figures expressed as a percentage of total hours worked, but not on the – what I consider critical – man-hours per sow basis. Critical, because how the sow performs is the basis of eventual productivity of the unit right through to shipping. "If you don't get 'em bred, you don't get 'em shipped," as one Iowan laconically remarked to me last year.

The total figures for each farm were taken off worksheets where available and the proportional splits agreed between staff and the management in each case.

Table 1. WORKLOAD EXPRESSED AS MAN-HOURS PER SOW PER YEAR

	40 farms 120-350 sows		10 farms 875-2040 sows	
Breeding to weaning:				
Feeding	4.2		2.1	
Serving	3.5		3.1	
Care and attention	2.5		1.8	
Moving	2.0		1.9	
Cleaning and disinfection	1.8		1.9	
Total breeding	14	50%	10.8	57.5%
Finishing:				
Feeding	4.5		1.2	
Moving and weighing	2.0		2.1	
Cleaning and disinfection	1.5		1.1	
Total finishing	8.0	30%	4.4	23.4%
Other tasks:				
Repairs and maintenance	2.6		2.1	
Records	1.1		0.8	
Other management	1.0		0.6	
Total other tasks	4.7	17%	3.5	18.6%
Building construction:	0.9	3%	0.1	0.5%
Total man-hours/sow/year	**27.6**	**100%**	**18.8**	**100%**
Finishing pigs produced/sow/year	19.8		20.1	
Liveweight produced/sow/year (kg)	1784		1850	
Labour cost/sow/year	€203.73		€187.53	
	(£138.50, $252.07)		(£127.57, $232.18)	

What these figures suggest...

• In farrow-to-finish labour cost terms, economy of scale reduced labour cost/sow by 8% (not as much as I expected).

- On the smaller units the amount of labour spent feeding is too high at 31.5% of total labour costs compared to 17.6% on the larger farms (largely due, in their case to automatic dry or CWF pipeline feeding systems).

- Conversely, cleaning and disinfection allowances of 14.5% and 16% of total labour input seem disproportionately low in these days of high disease risk. Is only one seventh of your labour effort enough? Spending six times more on labour for everything else these days when disease is our main threat to profit (after the fickle pig price), must be walking a tightrope, I guess?

- Note the typically high costs of moving pigs around (16.3% and 21.3%). This is one area where simplification, better housing design and automation will pay dividends in reducing labour demand. The situation is much worse on the larger units (as one would expect).

I think this is a fascinating survey, don't you?

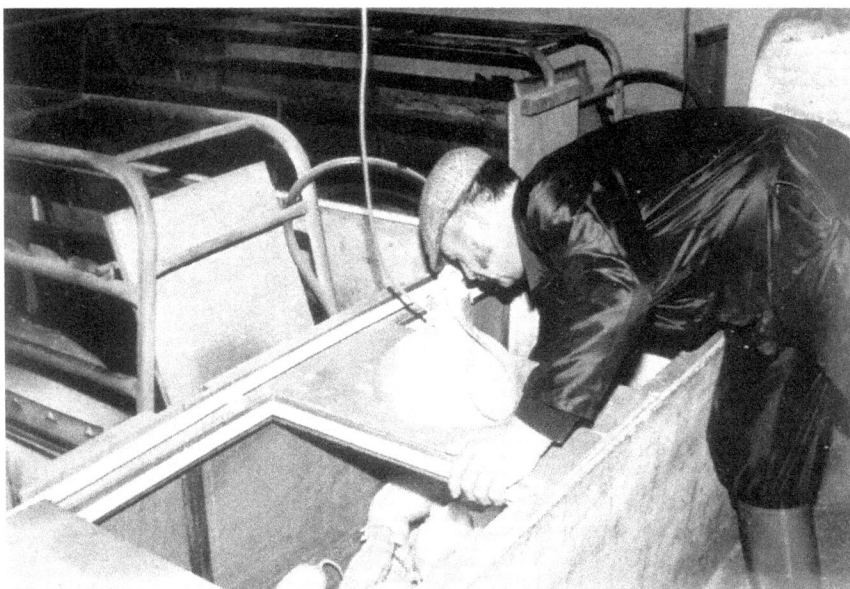

Pig technicians should be allowed adequate time to practise their undoubted stockmanship skills.

FEEDER ADJUSTMENT

This too is an important subject. I still find one in five of the self feed hoppers on the farms I tour to be incorrectly adjusted. Please, we must get it better!

Many producers are changing from dry free-access feeders to wet/dry single-space feeders and for weaners, dry pellet plate feeders. The reason – less food waste always seems to lead to better daily gain, usually lower FCR and more lean per tonne of feed. But often there is worse grading until the genetics are improved and/or the nutritional specs raised as the pigs seem to make better use of dampened feed. There is also less effluent produced – increasingly important these days.

I have surveyed 7 US trials and 11 European/Australian trials, all done versus conventional deep-trough dry feeders.

The results are consistent enough to show that when *properly adjusted* wet/dry feeders and plate feeders have shown a 0.15 to 0.28 FCR improvement against conventional dry feeders over 30-82 kg average growth period, and a 47-87 g/day improvement in growth rate, mostly due, I assume, to less feed being wasted.

However – on 3 out of 5 farms I visit there is evidence that feed delivery from wet/dry feeders is not well adjusted at all, and in fact on every farm I notice on my tour that there are always one or two feeders which are either showing too much or too little feed on the feedplate.

Some of this can be put down to the adjustment device being difficult to operate with the precision demanded, due to poor design. But the majority of cases are due to busy stockpeople *not checking the delivery of every single-space feeder every day*.

Except when the feeders are being introduced for the first time, when for 24 hours the gap should be raised to allow at least 5 mm (2") of pellets (50% more for meal) to flow forwards on the feedplate, as the amount of meal or pellets available should have to be 'worked-for' by a nibble or a lick. This gap (which varies according to the design of feeder used) has to be relatively narrow, so can get blocked. Thus constant vigilance, often using a wooden (or preferably rubber) truncheon to dislodge bridging, which also carries an L-shaped metal flange at the other end to free the gap of congealed feed, is vital.

It is a pity that no manufacturer has yet seen fit to market such a device which might help stockpeople to realise how important wet/dry feeder food delivery is to performance.

This is demonstrated by work done in Northern Ireland using different settings on a well-designed make of wet/dry feeder. Settings were adjusted to 1.4 grams, 2.7 grams and 5.3 grams per nudge of the feed release activator (*Table 1*).

Table 1 THE EFFECT OF FEED SETTINGS ON PIG PERFORMANCE

	Low	*Medium*	*High*
Food intake (kg/day)	1.97	2.14	2.21
Liveweight gain (g/day)	727	797	845
Food conversion efficiency *of carcass gain* (g/kg)	3.70	3.58	3.47
Backfat thickness at P_2 (mm)	10.6	11.1	12.1

From Walker & Morrow (1994)

The difference (estimates from the data) in saleable meat per tonne of food between the low and high settings was a staggering 29.5 kg, about as much as the cost of installing the device in the first place which could be recouped from the first batch of pigs!

Video recording was also used to examine behaviour (*Table 2*).

Table 2 THE EFFECT OF FEEDER SETTINGS ON PIG BEHAVIOR

	Low	*Medium*	*High*
N°. of feeder entries/pigs/24 hr	51.5	45.6	42.2
Feeding time/pig/24 hr (minutes)	110	78	87
Queuing incidents/pig/24 hr	70	45	26

Set too low to obtain the nutrient needs for the group, the Irish workers calculated that each pig would have to nudge the operating flaps an amazing and impossible 1400 times a day (about once every 5 seconds) compared with 420 times/day on the high setting. In practice the videos showed this increased their work rate by 165 per cent and they worked at feeding 26 percent longer.

Notice too that the number of queuing incidents – confrontations – was significantly less for the high group. This means less stress, itself a performance enhancer. While deep trough wet/dry feeders with flaps have largely disappeared in favour of circular plate type or shallow-trough feeders, the importance of correct settings on these newer devices is demonstrated by Table 3, carried out on a client's farm.

We deliberately set the flowgaps *too high*, ideal and too-restricted in comparison to Walker & Morrow's work where the settings were from ideal down to too-restricted.

Table 3 THE PENALTY OF INCORRECT FEEDER ADJUSTMENT (*PIGS 30 – 100 kg*)

Feeder gap :	Too restricted	Ideal	Too generous
Days to 100 kg	95.3	87.88	81.3
Food eaten/pig (kg)	190	188	207
MTF (kg)	270	279	248
Av Backfat at P_2 (mm)	10.4	11.0	13.2

Comment: The correct setting gave 31 kg more saleable lean per tonne of feed, worth €44.64 (£30.38, $55.27) at £1.10 ($2.00, €1.62)/kg deadweight pig price compared to a (careless) over-supply of pellets. In addition the number of downgraded pigs imposed a further income reduction (at 4.8 pigs/tonne of feed) of €1.31 (£0.89, $1.62)/pig overall, making a round figure of €61 (£41.50, $75.50)/tonne. Why express it this way? Because one feed dispenser should suffice, say, 3 batches of 15 pigs/year each eating about 200 kg/feed, or 9 tonnes of feed being dispensed per feeder/year. If the trouble taken to get the settings exactly right could be worth as much as €61/tonne additional income, then the cost of a feeder, about €192 (£130, $238), is paid back in under 4 months – again, as the Irish work suggests, on virtually the first batch of pigs put through.

Checklist on feeder adjustment

• When buying new single-space feeders for older growing pigs, or dry pellet plate feeders for nursery pigs, check the ease of feed flow alteration very carefully, including the *even* action of the restricter device across the delivery space.

• Special settings are needed after weaning. Depending on feeder type (consult the manufacturer), the following settings are guidelines for well-made small pellets:-

Weaning to 7 days after:

Allow two-thirds of the trough to be covered after a feeding session.

Weaning 7 days to 2 months after:

Reduce to one-third trough area covered.

To slaughter:

Never more than one quarter covered, less if possible.

- Obtain from the manufacturers their advice on ideal settings for age of pig and pellet size/meal volume.

- After a short run-in period to accustom new pigs (especially weaner pigs) to the source of feed, monitor the feed supply on to the plate daily. Settings will vary according to the design of the feeder, but in general the pigs should have to 'work' modestly for their feed, with a minimum amount of meal or pellets on free supply in the trough or on the plate. Seek advice from the manufacturer, as both over and under-supply can be costly in performance terms.

- Wet/dry meal feeders will require more food on to the mixer plate than dry plate feeders using pellets.

- Use a tool to dislodge bridging and clear feeder gap obstructions.

- Particle size – feed flow with meal is best at less than 700 microns.

- *Fill the hopper* before adjusting the flow space when starting from new after a cleandown. Do *not* fix the flow space with the hopper empty.

- Meal diets containing dried milk products tend to 'bridge' more than cereal and vegetable protein based diets.

CONDENSATION

If you are reading these words in winter, like as not you will be suffering from those annoying drips of water from metal transoms – even from the exposed nailheads used to fix insulation board to the ceilings. And plagued with damp areas above pig height when they should be dry!

Condensation is badly covered in all the housing textbooks I have on my shelves. Sure, some authors go into great detail, but they provide us with algebraic formulae and scientific diagrams which are of little help to the working farmer and his stockpeople.

What is condensation?

Water vapour always tries to move from an area of high humidity to one of lower humidity. As the vapour moves it cools down to the dewpoint and condenses into water inside the structure, often soaking the insulation layer. And that can reduce its heat-retention properties to almost nil as water conducts heat very well. The problem is noticed as wet patches on ceiling and walls.

Pigs are very wet animals. Quite apart from urine and dung, a pen of 12 midweights will exhale 5 litres (over a US gallon) of moisture a day. As this is in fine droplet form and warm, it rises upwards to condense on the nearest cool surface.

How to deal with condensation

• We all tend to put a damp-proof course under our concrete floors, but a damp course is equally important inside the building superstructure, especially if the building is pressure-ventilated either by positive or negative pressure. So, line the gap between facing-board and insulation layer with 1000-gauge polyethylene sheeting. This provides an effective vapour seal.

Next, we have to use airflow to keep us out of condensation trouble.

- Condensation occurs in 'dead spots' ventilation-wise, where the humid air has time to deposit its load of moisture droplets on a cold surface. It is important to know where the air bubbles are striking the surface causing the condensation, how fast the air is moving at that spot and from which direction it is coming.

- So know where your air currents are coming from, going to and where they are slowing down across a potentially cold surface. Use smoke tubes to ascertain this.

- Most stockpeople think that scouring the cold surface with a gentle airflow of 0.1 m/sec (about 10 seconds to cross a metre or yard) will remove condensation. It will help, but can be difficult to do, although judicious use of a 'table top' baffle board (see below) to deflect an air stream can help.

- Is there an air current passing fairly close to the condensation? If so, is it strong enough for some of it to be deflected to scour the area concerned? Use a simple deflector - a piece of plywood or a polyethylene sheet tacked to a light batten frame hung from plastic string at an angle sufficient to skid the air towards the desired route? Yes, it does look odd, collects dust and so forth - but it often works. It need not be permanent, and it costs nothing but a little patience in lengthening and shortening the string 'hangers' to get the angle of deflection right. Use your smoke tubes to check on what is happening.

- If the air pattern is too weak, you must consider increasing the fan power, sealing the building more thoroughly, reducing the inlet area nearest to the condensation or inserting an extra inlet nearer to the problem. Any of these solutions help. But at all times you should never accidentally increase the air flow over the pigs so that it causes a draught, unless it is summertime, of course, when they may need cooling.

- Coanda Deflection: The friction of a jet of air flowing along a smooth surface is less than the friction of the jet passing through the air. Odd, but true! Air entering a building may therefore follow or 'skid' along a smooth ceiling or wall far further than it would do across an open void. This is known as the Coanda effect. Look and see if there is a smooth flat surface near the condensation area and experiment by deflecting some incoming air onto the linking surface by using deflectors or altering the inlet configuration.

- Are any obstructions interfering with the Coanda effect? Can you remove them - pipes - battens - electric cables?

- A further extension of this can be to construct a duct, or cheaper, a polyethylene tube or tunnel, to blow cool air across a troublespot. Be careful about creating downdraughts at pig level, however.

- Insulation: If warmer, humid air condenses on a colder surface then it goes without saying that the problem can be cured by making that surface warmer by insulating it. I've left this until last because in most condensation cases this is either difficult, or it can be prohibitively expensive, to do. It should have been done earlier, but it hasn't. Sometimes however, surface insulation is the only answer. Some temporary insulation-board tacked on to a surface, spraying the troublespot with polyurethane foam and the use of "insulation paint" (for nail heads or small areas) are all solutions which have worked where stripping, re-insulation and re-panelling is ruled out.

Further tips to combat condensation

Inside ducts: Always divide the main internal air from the colder incoming air with 25 mm (1") of insulation board.

Trunk ventilators: Tend to condense towards the outer or upper lengths and must always be insulated all the way to stop dripback.

Encased Natural Ventilation exits: Sometimes internal air is significantly slowed from rising, or stopped altogether in still air conditions on the outside by a 'plug' of cold, heavy air once the exit leaves the roof, particularly if the exits are round metal ventilators. Any protruding trunking must be well-insulated too.

Drop-out fail safe panels: these often condense on the upper surface. The panel should be made of insulation board. They rarely are.

Sometimes surface insulation is the only answer

To help with the technical terms involved, the table overleaf could be useful.

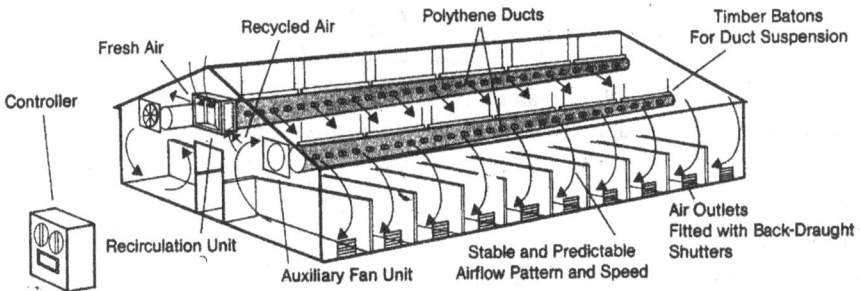

Figure 1. The 'Hybrid-Recirc' ventilation system. This airbag principle can stop condensation.

Table 1   GRAPHICAL EXPLANATION OF INSULATION TERMS

K Value = Thermal Conductivity of a Material

Measured in: Watts* per metre per °C (W/m °C)

Thermal conductivity. A measure of a material's ability to conduct heat. Defined as the quantity of heat (watts) which will flow between two opposite faces of a 1m² of material with a 1°C temperature difference between those two opposite faces.

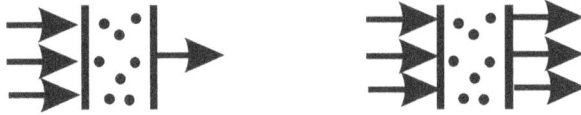

(Different materials)
Note: The materials are the same thickness

Lowest values are best. Typical insulation materials = 0.02 to 0.2 (straw = 0.07, copper wire = 200)

R Value = Thermal Resistance of a Material

Measured in: Square metres per °C per watt (M² °C/W)

Thermal resistance. A measure of the resistance to heat flow of 1m² of a given thickness of material, or structural component made up of a number of materials.

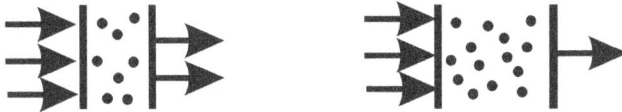

(Same material)
Note: Different thicknesses of the same material

Highest values are best: **Range**: 0.12 to 0.55 for typical piggery measurements/materials

U Value = Thermal transmittance of a structure i.e. can be a combination of various materials

Measured in: Watts per square metre per °C (W/m² °C)

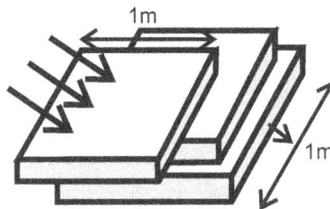

Note: K value (above) refers to an **individual** material

Note: U value is the amount of heat which passes through 1 square metre of the materials comprising the whole structure when the temperature difference from outside to inside is 1°C.

Lowest values are best. **Typical range**: 0.5 to 5.5

What is a Watt?

The basic metric unit of energy is the Joule (J) which is used to measure various forms of energy, including heat energy. 1 joule = 0.239 calories. Heat flow is the rate of transmission of heat. The metric unit of heat flow is the joule per second, called the watt (W).

MOVING AND HANDLING PIGS FOR SHIPPING

No textbook I've read has covered this commonplace task anything like well enough. It is such a routine job, done so frequently that all of us – myself included – have tended to shove it to the back of our minds. In fact there is quite a bit of information on the subject thanks to pioneer researchers and animal behaviourists like Prof. Temple Grandin, Drs Pedersen, Gonyou, Guise, McGlone and others.

Moving incalcitrant pigs can be frustrating to them and to stockpeople. Anything which raises stress these days is bad for the pigs, bad for the processor – poorer meat quality – and not-so-good for pig technicians either!

Here are some points I've picked up from scanning 20 or more papers on the subject. They are all interesting but not in any order of importance – just listed in the order I came across them.

Two types

There are definitely genetically nervous pigs and calmer pigs. To make handling easier, differing techniques are advised.

- *Nervous pigs*. These need quite a bit of **daily** movement among them by the stockperson. It is important to **keep moving at walking pace** to accustom such pigs to flow round and past you. 10-15 seconds per day for 50 pigs is advised as a minimum. But just doing this for a week prior to shipping is counter-productive as come shipping date they are **more** nervous. You've left it far to late. Walking through 'Big-Pen' systems and/ or the 'Wean-to-Finish' buildings is especially valuable. Making time to do this, and observing as you go, has to be part of our stockmanship skills nowadays.

- *More docile pigs*. Here one must not spend too much time in the pens; doing so makes the pigs so tame that they resent being driven. So look at them 'over the gate' so to speak.

Footnote: Having written these thoughts out, I've just turned to what I consider to be the best textbook on pig production ever written to see what the authors thought. I could find no mention of moving or handling pigs! Enough said?

Happy medium

- The animal needs to be docile enough not to panic when the time comes to be driven, but not so tame as to hang back – a desire to follow you rather than be driven on ahead. There is a happy medium in the time you spend walking among the pigs. Remember – keep walking – no boot chewing! Pigs need to get used to flowing past you.

- Alleyways must be 1 metre wide – this allows two pigs to walk side by side. No pig likes to be the leader, but if one is curious this will slow everyone up as it 'dithers'/explores. Two pigs together give each other confidence and some competition in curiosity. If passageways are narrower than this, only 3 pigs moved at one time is advisable, not a whole truckload. Anyway, research suggests moving 5 or 6 pigs at a time gives least trouble; lowers the stress reaction. But big units will laugh at this advice, not surprisingly!

- Never stockpile pigs in an aisle before driving – this really gets them stirred up/nervous that something is about to happen.

- If you can spare the time (not to mention the cleaning up afterwards) it is a good idea to allow the pigs to walk the access passage every day – or as often as you can fit it in. But on a busy farm this is not very practical advice either, even if the research suggests it.

Drafting / sorting

- Keep calm, move slowly, **never** use electric prods, slap with sticks, shout or bang gates. Stirred-up, excited pigs will stick together for comfort and be more difficult to separate.

- The sorting board remains the best device when pigs have to be moved, so latest research (McGlone, 2004) reveals.

- When exiting a building pigs do not like air blowing onto their faces, so stop the ventilation for a while or pressurise the structure temporarily so that the airflow rides with them.

- Pigs are hesitant to move from a dark surface across a distinctly lighter one – even a narrow bar or a puddle reflecting light. The converse can also sometimes cause hesitancy – pigs don't like entering a dark area – so light the exit well.

- A good tip from behaviourist Grandin is "To overcome that first reluctance of pigs to exit the building, attach 5 m of plywood to the pens nearest the door. This will prevent pigs that are being driven from seeing or touching pigs that are in pens near the door. Remove the plywood once they are loaded." Why near the exit door? It seems that once the possibility of exiting the familiar building arrives the pigs' desire to 'fraternise' goes up a notch, so holding things up. I've seen this quite often myself.

- Avoid having anyone visible up-ahead. So many times I have seen the truck driver standing by the vehicle ramp to assist with loading – he thinks helpfully. In fact he is a distraction and could even be considered a 'threat' – so keep him out of sight. In fact, biosecurity-wise, ***never let him anywhere inside the pig premises either!*** He is a major disease threat.

- Loading in the winter is often done in early morning when it is dark; light the route to the tailgate and shine a light from ***behind the pigs*** into the truck bed. It is amazing how this reduces reluctance to climb the slope up to it.

- If you back a truck up to a loading chute, if possible there should be slight step ***down*** for the pigs to take into the truck, not a step ***up***, however slight. As truck beds vary in height it is a good idea to have at least one metre of hinged ribbed flap to ameliorate any excessive drop down. Cover it with a *little* bedding.

- Many loading chutes have a funnel-shaped exit into the transfer race, which can cause pigs to jam. Making the exit offset (see Figure 1) removes most of this nuisance. Installing a vertical roller helps as well.

- Gathering / assembly yards ('crowd pens' in North America) often have right angle corners. Better at the design stage, is to plan for curved or obtuse-angled solid sides, at least 1.2m high. Groups will often face inwards at right angles and are then laborious to un-jam. I find 25 individuals are enough in any crowd pen.

- In localities where timber is plentiful, loading ramps are made of wood, which very soon becomes slippery. Raised wooden cleats should be used (in my experience non-slip compounds are soon worn away and cut-in grooves aren't good enough). Cleats should be at least 25 mm high and 200 mm apart which will suffice all sorts, including sows. ***But watch wear*** from protruding nails!

  The slope of a loading ramp should not exceed 20 degrees from the horizontal (Figure 3).

An ingenious idea . . .

- Figure 2 overleaf shows a cross-section of a novel race suggested by Dr Grandin as long ago as 1982. The narrowing of the tread channel to 18 cm seems to encourage the pigs to move forward – almost as if the slightly-constricted placement of the feet gives them a feeling of 'falling forwards'. Also the side-by-side design encourages following behaviour. I've only seen one example of this ingenious design, and this was a larger version for weaned sows, stretching over 100m or more, and the way the sows, when left to themselves, moved along rather like being on a conveyor belt was both impressive – and rather amusing – to watch, but I had no video camera with me!

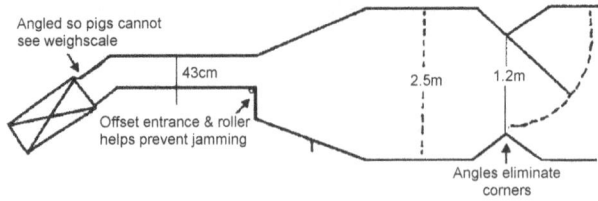

Figure 1. Treatment/weighscales chute with entrance to prevent jamming at the chute entrance. The angled fences at the rear gate eliminate corners in which the pigs could bunch up. The exit chute Is angled to prevent incoming pigs from seeing the scales and balking. Source: Grandin (1982).

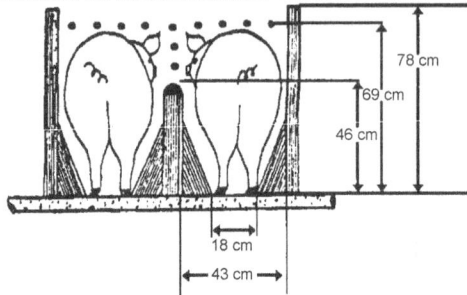

Figure 2. Two chutes side by side are better than one. The outer sides should be solid to prevent the pigs from seeing distractions outside the chute. The centre partition should be constructed so the pigs can see each other in the adjacent chute. This encourages following behaviour and improves the flow of hogs. Source: Grandin (1982).

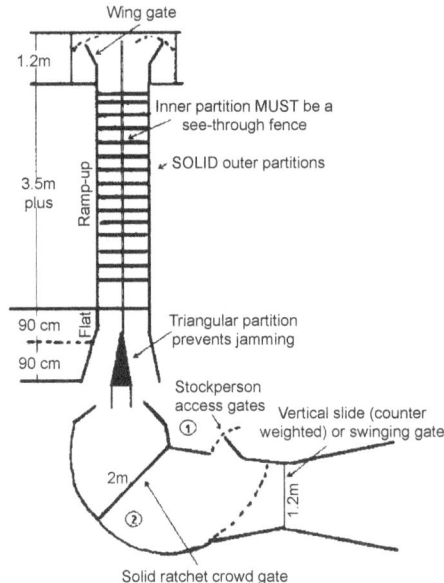

Figure 3. A loading ramp with two single-file chutes. The handler stands at ① when the crowding pen is full and directs the leaders into the chutes. As the pen empties, the handler steps through the gate into position ② and swings the gate around.

Reference

McGlone *et al.* (2004) 'Pig-Moving Devices Tested for Effectiveness'. Nat. Hog Farmer Swine Research Review, Dec. 2004.

DOING A STRESS AUDIT

Open almost any pig textbook and you will find the subject of stress badly covered. Mention that to a pig scientist and he will reply, "That figures – stress is complex and is almost impossible to quantify. Any reliable conclusions / actions to be drawn from an attempt at a scientific approach towards methods of mitigating stress must start from a more reliable method of measuring the various forms of stress in animals than we presently possess."

All very true – if looked at from the stance of science back down to the animal. Now turn the problem the other way round. Approach it from what the animal – in our case the pig – tells you, or is seeming to tell you. Then see if you can identify, put right or ameliorate the visible, audible and physiological clues which are often directed at you, the owner/stockperson, but not picked up nearly often enough.

Stress - not detecting the signals

Pigs talk to us all the time. Vocally (a little). Body language (a lot). Change in habit/routine (considerable). Inter-reactive behaviour (even more). As humans, we are quite quick to detect stress in those close to us and other humans we meet during the course of the day, so why are we so slow in picking up the similar signals in our pigs?

A case of over-familiarity breeding – not contempt, but lack of response, I guess. Again, we are often so busy with the job in hand, or the next one beckoning, that we don't really LOOK at pigs!

Some scientists are quite convinced that stress damages immune response, and that part of the reason why certain diseases seem to be winning hands down just now is that the pig's *natural* defences to them are compromised and that stress plays its part in this.

The stress audit

I couldn't agree more. I've been walking pig units for 40 years and had a hand in managing pig farms on and off during that time. For 30 years I've encouraged and trained pig producers and farm students to do a pig stress audit.

Here are some tips.

A periodic stress audit is a very useful exercise, probably needed every 6 months on many units. The purpose is threefold.

1)   To identify, from the body language of the pigs before you, their responses to the likely stressors they meet.

2)   To check that the conditions you have imposed on the pigs are suitable and that the equipment and management impositions on the animals are up to standard and within acceptable boundaries.

3)   To work out and provide instructions for their rectification if any shortfall is found wanting or approaching borderline.

A few experiences from 30 years of stress auditing

•   Do it quietly! Observe the pigs under their normal behaviour patterns. Open nursery doors an inch, and listen carefully before switching on the light and/or entering. Listen to the breathing, for restlessness, wheezing, 'snittening' (light, irritant sneezing). The same in the farrowing house and in the grower houses.

•   Observe them unawares. Still with the door slightly open, switch on the light and take time to observe lying pattern, distance apart from each other, where they are lying in relation to pen furniture, floor type and ventilation pattern. Good observation of pen behaviour can pick up major departures from this routine.

When I do a stress audit for a client, this is what I ask myself…

•   Are his pigs within the published temperature comfort zone?

•   Is he frustrating the pig in any way? Food intake, and quality?

•   Access to water and intake (flow-rate). Exercise; including room to flee/ avoid aggression.

•   Is he counteracting aggression? Is stocking density and pen shape correct? Too many sows together? Poor trough design? It seems extraordinary to me that while recent work has shown quite clearly that 'winged' feeding spaces reduce aggressive incidents, few people have taken up this cheap and simple idea (see page 69). Then is there poor drinker siting? Many people site drinkers in corners, the one place they should not be! A pig is trapped when *facing inwards* drinking in a corner. But not nearly so much when defecating, which is why they chose a corner to do this. *Facing outwards* there is good protection from the corner against being aggressed from behind or from the sides during a critical natural function.

• Is the floor type stressful? In the interests of easy cleaning many are – the in-contact area is too small for tiny feet. If so, does he provide a solid 'comfort board' for tenderfooted weaners so that they can at least get on to a solid area as a respite?

• Is the air quality good enough? It is simple to measure gases and dust accumulations and clear maximum standards exist. Is he doing so?

• Are the pigs quiet? If not, *where* is the focus of disturbance, and *why* is it there?

Stress, strain and pain

Stress: In general is looked upon as a symptom resulting from exposure of an animal to a hostile environment. ("Environment" in this context refers to the conditions under which the pig lives and which affect its behaviour, not just its housing). So more accurately, stress relates to any external forces which disturb the pig's perception of normality, which the scientists call 'homeostasis'.

Strain: Is the internal displacement brought about by stress; the results of being in a stressful state. Strain is caused by stress.

Failing to distinguish between stress and strain causes confusion when scientists try to measure stress. The degree of environmental stress can only be measured indirectly through the response of the animal. Thus we measure *strain* (in a variety of ways, but mainly using strain indicators, primarily by measuring hormone changes in the body, but also to some extent organ-specific enzyme and haematological – blood – changes) so as to try to assess the degree of impact of the stressors which cause strain.

Even today, and especially with pigs, we know too little about the mechanisms involved and what the animal considers 'normal' i.e. unstressful. Here research *is* needed, and the world-wide move to improve animal welfare is – at last – gradually providing this.

Pain: We all know that pain is stressful. Animal behaviourists believe that alert, aware animals can have a similar experience of suffering if maybe not of the same complexity, as human pain. While attempts are now being considered to use behavioural and physiological changes to assess the degree of pain which results from routine husbandry practices such as tail-docking, castration, ear-notching and slap marking – or even teeth-clipping, it will be some time before this is available, if parallel work on the rubber ring tail-docking and castration of lambs is anything to go by. It is slow, meticulous and expensive work, but at least it shows promise in being able to rank pain in relation to various bodily and behavioural responses.

There are many, many things we can do to reduce stress and strain and so help boost natural immunity to disease. On every farm that I do a stress audit, I find there are a minimum of five probable causes to be discussed: satiety; flooring and comfort; companion interaction; pattern of routine; boredom; and a further five ideas which could help reduce it – from dietary redesign / supplements / to ventilation adjustment, water manipulation through to alterations in management.

The fact that many scientists think the area is too imprecise to make best use of their limited financial resources for research into stress is their business. At the sharp end we can do much ourselves until they can get round to helping us through the scientific approach.

Finally, a few simple things you could think about.

- If pigs talk to you, should we talk to them? Why not? At worst, it helps pass the hours!
- Playing music? The familiarity of background sound must reassure when quarters are changed and pigs are mixed.
- Sticking to a time routine? All confined animals habituate and going along with their expectations could lower stress.
- Give them toys? Sure, why not? They must get bored stiff!
- Get someone else in? Sure, one of the best people to ask in occasionally is another experienced breeder, nurseryman or head stockperson. He will see what you don't due to your familiarity with your own unit.

   Danger of him bringing his diseases with him? Not so! Read page 192.
- If you can, do think about housing sows in deep straw! It simply melts away stress, designed and managed well.

Of course, it is dangerous to become anthropomorphic and assume that pigs respond to what we humans prefer. This said, why do the best pig stockpeople use anthropomorphism more than they care to admit?

It helps to lower stress – on both sides.

WATER – THE BASICS

At last modern textbooks cover water well, but the findings are somewhat scattered. I will try to draw some of the basics together. At the same time, over a period of 15 years, I was working on pig units in tropical and sub-tropical countries, both hot-wet (Philippines, Thailand, Vietnam, parts of SW China, Okinawa, Kyushu – Japan) and hot-dry (Australia, Mexico and Spain). Watering pigs comes top of the list in all these countries and is generally plentiful. As a result I learned a lot about water.

Most pig producers believe that the pig drinks water to process its food and, given unrestrained access to it, will drink its fill to satisfy this need and then instinctively stop.

Water intake is more complicated than this.

Why does the pig drink?

1.   Certainly, to facilitate its metabolic processes….

    •   Transport of nutrients – and hormones – through the body.

    •   Maintain a correct acid/alkali balance in the body.

    •   Process minerals.

    •   Remove the end-products of normal digestion, especially protein residues, and certain minerals in excess, like salt.

    •   Detoxify its body when necessary, ie remove drugs, drug residues and toxins.

But also….

2.   For thermo-regulatory purposes. To adjust and help control body temperature.

3.   Help achieve gut fill and enjoy the well-being of satiation.

4.   Mitigate/enhance behaviour, i.e. boredom, entertainment, devilment.

Rather more is involved than just processing food and slaking its thirst!

Is water still the "forgotten nutrient"?

Not in the nutritionists' world. Several researchers have done excellent work and academically water is no longer ignored. With many pig farmers, however, water still has a low priority – "Just make sure the drinkers are working properly at all times" is thought to be enough. It isn't in these days of hairline profits.

I believe farmers still need to catch up on the recent research on water – admirably summarized by Prof. Peter Brooks of Plymouth University, England, ten years ago during his team's pioneering work on water for pigs, and his conclusions have never been bettered. I list his views below, adding a few comments of my own from on-farm experience.

Brooks: "Where water availability is limited (irrespective of the reason for limitation) feed intake is restricted. The animal will not consume food, the waste products from the metabolism of which it is unable to excrete.

Comment: This is particularly important where scouring, especially post-weaning diarrhoea, occurs. The scouring piglet very quickly uses up the reserve of water in its body tissues to 'flush' the harmful bacteria and their toxins safely down the gut to the outside. If it cannot take in (*insorb*) enough replacement water through the gut wall, it steadily starts to poison itself, being unable to remove sufficient end-products of metabolism from its body. At the same time the blood thickens, delaying or preventing sufficient oxygen being transported to the muscle and surface tissues, so the piglet feels cold, 'tired' and miserable.

This is why, as soon as digestive looseness starts, ample clean water must be available, preferably treated with electrolytes. These enable the young pig to take in fresh supplies of water *while at the same time* allowing the gut wall to pass out (*exsorb*) toxins. This is an essential part of water management. At the time of writing two thirds of those farms I visit haven't the electrolyte stock solution ready to hand at the first sign of looseness.

Brooks: "As water availability increases, voluntary feed intake will increase proportionally up to a maximum set by the total volumetric intake limit."

Comment: This is why, in order not to limit the pig's ability to 'drink its fill', we need two drinkers in each pen; why sows should have a bowl/trough waterer in lactation, not an awkwardly-sited bite drinker; why those Japanese cup/leaf drinkers are so good for sucklers; why weaners should be encouraged to drink water before and after the weaning process, etc.

Brooks: "Where feed intake is limited and water availability is not, the pig will increase water intake to satisfy its demand for gut fill."

Comment: No-one deliberately underfeeds a pig these days – or do they? But feed limitation is quite commonly seen in gestating sows, where a touch of fibrous food, e.g. dried sugar beet pulp (but not too much of it) will help satisfy hunger at a time when the sow is largely on 'nutritional' holiday, anyway, and being cut back in energy.

Brooks: "Where water availability (consumption) is adversely affected by palatability, the problem may be rectified by flavouring the water."

Comment: This, in my experience, is quite a tricky area! The first remedial action should be to remove the source of unpalatability in the water supply, but in parts of Canada and Australia, where water has to be obtained from deep-ground aquifers, the trace-element contaminants are irremovable (apart from expensive filtration which is not always possible) and it appears these are extremely difficult to mask with a flavourant.

When we tried adding flavourings which we knew the pigs relished (from adding them to feed), this seemed to have an opposite effect when they were added to the highly-mineralised water – at varying levels – apparently making the water more unpalatable, as the pigs reverted to higher consumption when we removed them from the water.

My advice is to proceed cautiously on this one. Flavouring water is worth a try, but be prepared for disappointment. Brooks does say "may be rectified".

Brooks: "Where water flavourings are particularly attractive, consumption may be stimulated beyond the normal intake. This can result in water comprising an abnormal proportion of the total volumetric fill. In such cases feed intake will be decreased."

Comment: The same occurs in hot dry countries where heat causes over-drinking. The only solution is to spray-cool the pigs. With global warming in the colder pig keeping areas of the world (e.g. Mid-West USA, Ontario Canada, NW Europe and perhaps Brittany, France and Catalonia, Spain) all meteorologically said to be moving to the warmer south at up to 20 metres a day, pig producers in these areas would do well to bone up on the whole well-researched area of spray-cooling equipment and the water allowances advised to keep pigs cooler.

Maybe not all the reduction in feed intake is due also directly to heat stultifying appetite; perhaps higher water intake is also to blame?

Insufficient access to water

Common errors are not having at least 2 drinkers per pen of more than 15 pigs and too weak a supply at the drinker (see Drinkers and Flow Rate p. 149). Also underwatering due to awkward access in the farrowing crate. Post-farrowing dehydration and subsequent lactation milk formation demand is extremely thirst-making.

Where weaners are concerned, increasing drinker flow rate from 200 ml/minute to 450 ml/minute increased daily gain by 30g/day (weaners 7-25 kg) on one client's farm. Peter Brooks also reports that increasing flow rate from 200 to 400 ml/minute among 3 go 6 week weaned pigs increased the time spent drinking from 3 minutes to 5 minutes per day. Because of the lower water intake, the low flow rate weaners ate less food.

Reference

Brooks, P 'Why Worry About Water'; (n.d.) Monograph issued by University of Plymouth, Faculty of Agr. Food and Land Use.

TWO GOOD NEW IDEAS WITH WATER

Being a wanderer across the world on pig business for 20 or more years, I come across many good ideas. Some are taken up avidly by producers, but other ideas – for what seems to me to be unfathomable reasons – have never really caught on, or are being adopted too slowly for my impatient and enthusiastic judgement!

Things like Pig Train – an obvious development in order to minimise any moving or mixing stress in the growing pig. I've measured moving and mixing stress on clients' farms when a no-move opportunity arose. Every move cost them two days growth. The Pig Train has gone – until someone re-discovers it! Like the Japanese pipe-house. This too has gone – but not quite lost, as the Australian Eco-shelter achieves half of its purpose but doesn't necessarily copy its brilliant composting concept. That part of it will dawn on us as soon as eco-friendly manure handling restrictions bite further into our freedom to farm as we want to. Ask the Dutch!

Like pipeline pig feeding, which as taken 35 years to get really taken up (except much of N. America, who are still 30 years adrift on this concept, and as a European, I'm quite happy that they are!) Like Blend Feeding, which when taken together with the Challenge Feeding concept opens the door to several new and probably more efficient ways of feeding the growing pig – and moreover, could mean the end of the costly animal feed salesman – should the global feed compounding industry wake up to how much it could save them!

Like…. No, I must stop! So many good ideas which have bitten the dust at least for the time being. But I'd like to put before you, again, an article on this theme I wrote 8 years ago. Not a whisper on either idea has appeared in any textbook yet, or from the serious animal researchers.

Twenty years ago those of us consultants who work on farms considered ourselves ahead of the academics on the subject of watering pigs. Yours truly campaigned in the 70's and 80's for more scientific work on water – at that time there were but 15 papers in my files on the subject. Now, thankfully, there are 47. I'm told there exist at least 100!

But despite all this recent valuable and worthy work, we sharp-end guys could still be slightly ahead of the academic pack! Let me give you two examples of how we farm advisers have been trying out new concepts which – as far as I can fathom in visiting a dozen pig industries each year – the scientists haven't yet explored.

Oasis drinkers

I and my clients have been using this idea for several years now. The concept was copied from the turkey boys and now uses an adapted plastic dome-shaped turkey drinker for weaner pigs.

I'm sure once you've tried it you'll never go back to conventional bite drinkers again.

And yet they haven't really caught on worldwide as I think they ought to, something which has puzzled me.

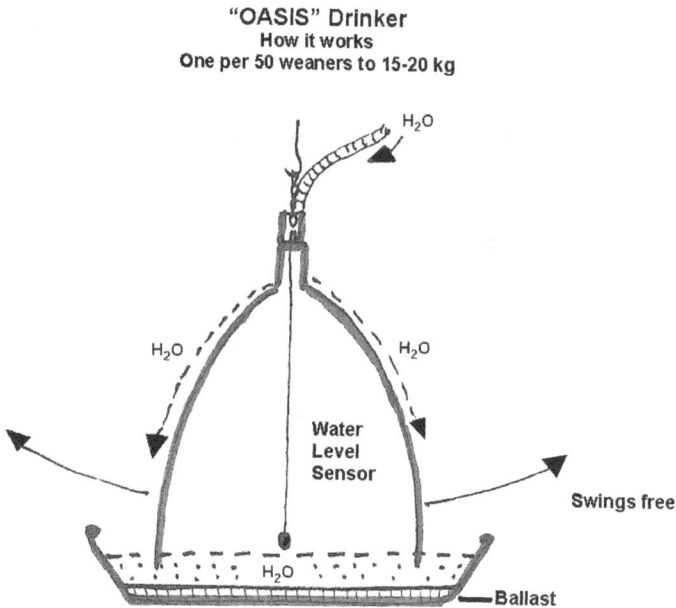

"OASIS" Drinker
How it works
One per 50 weaners to 15-20 kg

Figure 1.

My diagram (*Figure 1*) describes the principle which is a suspended plastic dome down which water flows into a circular trough when a 'top-up' switch is activated by the water level falling as pigs drink. The device is ballasted to allow a slight swing and to prevent upsetting. They are not expensive and easily moved from pen to pen or within a pen.

The slight swinging/rotation action plus the flow of the water down the sides of the dome seems to attract the pigs first to lick the surface, then drink from the trough.

It is claimed that weaners especially do much better on this device than through either bite or nipple drinkers (2 per pen).

We tested them out on 500 weaners at 20 per drinker against a similar number of controls on conventional bite drinkers.

The results were : –

Table 1. TURKEY DOME DRINKERS V. CONVENTIONAL DRINKERS

	Oasis	*Bite drinkers*
Weight in (kg)	6.3	6.3
Weight out (kg)	24.2	21.7
Days on trial	41	41
Av daily gain (g)	450.0	390.0
Food eaten pig (kg)	66.75	62.05
FCR	1.73	1.77
Liveweight per tonne of feed (kg)	578.0	565.0

The bottom line

At the end of the nursery stage the Oasis-watered pigs provided an extra 13 kg of weight per tonne fed. As one tonne of feed suffices about 30 pigs, this would need 1½ drinkers costing £40 in the UK or plus 10% interest (reducing) over a 5 year life – say £55 ($100, €80).

The 13 kg more weight per 30 pigs/1 tonne feed is worth £15 ($27.30, €22.05) thus the payback on the basis of this trial is 55 ÷ 15 = 3.7 years at the end of the nursery stage. A bit long? Could be.

However . . .

We followed the pigs through to slaughter as pig producers sell finished pig meat, not weaners but both groups were now put on conventional waterers.

Table 2. TRIAL TO CHECK ANY 'CARRYOVER EFFECT' TO SLAUGHTER

	Weaner-stage Oasis drinkers	*Weaner-stage Bite drinkers*
Weight in (kg)	24.2	21.7
Weight out (kg)	90.4	90.0
Av. daily gain (g)	818.00	809.00
F.C.R.	2.41	2.48
K.O.%	74.20	73.90
Saleable meat per tonne of feed (kg)	420.0	398.0

The carry-over effect seems to have got 22 kg more saleable meat per tonne fed, and so the payback shortens to 1.6 years. This is much better.

'Elephant troughs'

Sows **hurt** after farrowing – ask any woman after childbirth! The last thing she wants to do is get up and sit down frequently, at least for a day or two.

Yet what do we thick-headed designers (exclusively men!) of farrowing crates do? We put silly little bite or nose-press drinkers, often set at far too low an angle which even at a 1.5 litre/minute flow rate is far too uncomfortable for a sow which may need to drink 8 gallons a day; more if it's hot. Try drinking a can of Coke at the level of your knees a day after a hernia operation – see what I'm driving at?

No, what the sow wants is to get up and put her snout into a nice 8-litre pool of water and vacuum it up like an elephant – then sit down with a sigh or relief! You watch them if you do this – that's just what they do !

PROOF

Many years ago on our farm at Dean's Grove, Dorset, England, we did this. We provided what we called 'Elephant Troughs'! We never measured intake or food consumption, but Gordon, our stockman, was certain it was beneficial. Even so, this wasn't proof.

So in several cases of Far Eastern sows failing to eat enough in hot weather I got them to change to water troughs alongside the feed trough. The difference – from a poor starting base it is true – was dramatic (*Table 3*). One farm then did a side-by-side trial before making more alterations having also incorporated several appetite stimulatory ideas of mine. Even then the increase in water intake – yes probably with more water wastage too – was noticeable if not statistically significant, I'm told, due to the small numbers compared. But the acid test was that still they ate even more in lactation (*Table 4*) and productivity got a further boost.

So, my academic friends – here's two practical ideas you may like to investigate! All I can offer at present is specific farm trial data – we need your harder evidence to back up what I'm sure are two very good practical ideas with water.

Table 3. PROVIDING WATER IN 2 GALLON TROUGHS IN PLACE OF GOOD QUALITY BITE DRINKERS (TROPICS)

Farm	Average lactation feed intake day kg (lb)			
	Before		After	
1	4.7	(10.6)	6.4	(14.1)
2	5.0	(11.0)	5.9	(13.0)
3	4.1 or less	(9.0)	'6.0 or more'	(13.2)

Table 4. BREEDING FARM (TROPICS) WITH LACTATION APPETITE PROBLEMS, PART-RESPONSIBLE FOR POOR PRODUCTIVITY

	Initial	*After 5 management alterations; bite drinkers still retained*	*After, but also with drinkers replaced with troughs*
Average lactation feed intake day kg (lb)	4.17 (9.38)	5.03 (11.09)	5.51 (12.15)
N° of sows	426	432	24
Av. Weaned per litter kg (lb)	40.1 (88.4)	46.4 (102.3)	47.87 (105.6)
Farrowing Index	1.93	2.11	2.19
Av weaning wt per sow per year kg (lb)	77.39 (170.6)	97.9 (215.9)	104.8 (231.1)
Weaner wt per tonne feed kg (lb)	76.6 (168.9)	84.4 (186.1)	87.3 (192.5)

The Bottom Line Across the whole herd alterations improved productivity per sow by 26% and productivity per tonne of feed by 10%. This rose to +35% and +14% respectively from the sows in one farrowing house which also had access to 'elephant troughs' instead of bite drinkers. But note the low initial productivity starting base.

A plastic turkey drinker is adapted as a waterer for weaned pigs.

Table 5 COMPARISON OF CONSUMPTION/PERFORMANCE OF LACTATING SOWS WHICH WERE TROUGH-WATERED V. DRINKER-WATERED, SUMMER CONDITIONS, FIRST 21 DAYS OF LACTATION

	Trough	*Drinker*
Water used/day, litres	23	18
Water spillage, litres	not recorded	3.6
Food eaten to 21 days, kg	139	127
Piglet weight at weaning, kg	6.03	5.81
Piglet growth rate/day, grammes	223	212

Source: Clients' records (USA)

Table 6 WASTE OF WATER FROM THE SAME MAKE OF DRINKERS SET HORIZONTALLY TO ONE SIDE AT 11cm (4") FROM THE FLOOR AND AN ADJUSTABLE DRINKER SET AT 15° AND 4" ABOVE THE SOW'S BACK-LINE

	Water used/day (litres)	*Lactation feed eaten 0-21 days (kg)*
Low-set	25.0 (+42%)	122
Correctly-set	17.6	136

Source: Clients' records (USA)

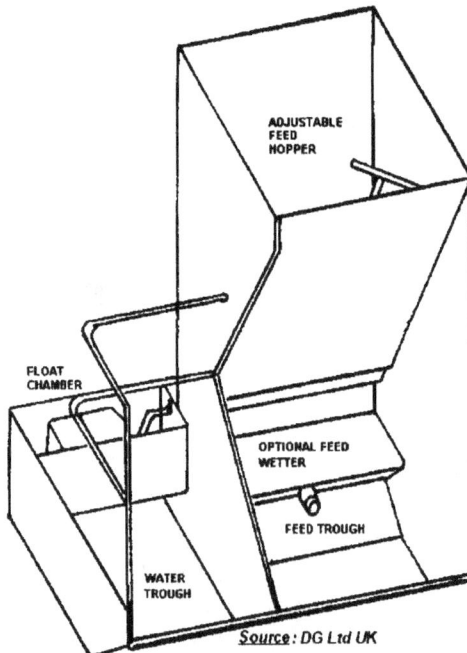

A good sow lactation feeding station incorporating an 'Elephant Trough'

MORE ON WATER.
REQUIREMENTS, FLOW RATE AND SOME
THOUGHTS ON DRINKERS

Most textbooks on pigs provide 'helpful' tables on the water requirements of pigs and Table 1 is typical. I say helpful because many good scientists have spent time and effort in compiling them. But *are* they all that helpful? I think not.

Table 1 WATER REQUIREMENT OF PIGS (A TYPICAL TEXTBOOK TABLE)

Pigs weight	Litres per day
(suckler)	0.27 (mostly sow's milk)
Piglets to 15 kg	1.2
15-40 kg	2.25
40 – 60 kg	5.0
60 kg plus	6.0
Dry sow	5.0
Pregnant sow	8.0
Lactating sow	15-20
Boar	9-11

How *can* this sort of thing be accurate? At best it is a very rough stab at it, useful mainly for checking on header tank capacity. I have growing evidence that farmers and manufactures are paying too much lip-service to theoretical daily needs, which are hugely variable from week-to-week and often from day-to-day, which leads them to ignore two most important things. The following factors can, or do, play major parts in making a pig more – or less – thirsty.

•   Temperature, relative humidity (dryness), dustiness of, and speed of, air.

•   Mineral content, texture and dustiness of feed (including what binders are used in pellets/nuts).

- Stocking density in relation to drinker siting, height and flow rate.

- What feed raw materials are used.

- Disease level, including mycotoxins in feed.

- Boredom / stereotypic behaviour.

- In sows, litter size/demand for milk.

So such tables are not sacrosanct – only a guide. Variation can be between 40-60 per cent in cool weather alone. ***Pigs should be given as much water as they want***, and this makes flow rate and header tank capacity (if not watered directly off the mains supply) the two bottlenecks. Table 2 gives a guideline for header tank allowance.

Table 2. HEADER TANK ADEQUACY

Assume a safety factor of 500 x 50 kg pigs in 20s, needing 2500 litres per day under normal conditions. Generally speaking, the peak demand safety allowance needs to be 25 to 33 percent of daily needs, ie in this case 650 to 800 litres which calls for 7 to 9 x 100 litre header tanks sited down the building. Few people have this capacity. With 500 pigs drinking/wasting water at only 250 ml per minute (minimal flow-rate) and with only 20 percent occupancy of 50 drinkers at any one time, a 100-litre header tank will run dry in 30 minutes and a 50 litre tank in only 15 minutes.

Flow rates

You will see varying advice on flow rates in the textbooks. This is because there is a range of factors affecting the subject...

- Pigs tend to have different drinking speeds.

- Over-generous flow rates waste water, which is expensive to remove as a component of slurry.

- Too high a head of pressure can dissuade weaners from drinking, at least for a day or two after weaning.

- A higher replenishment rate is required for older pigs, especially lactating sows when watered by bowls or a trough (better) than from bite or nose press (the latter not advised) drinker valves. A sow can drink much faster from a pool of water than from a drinker.

- Header tanks can run low/dry.

The following guidelines are given from my own experience and that of my clients who have helped, together with what I consider to be the best of the research so far published.

Let us look first at drinking speed, which as I've said, is variable, but the research suggests ...

Table 3. TIME TO DRINK ONE LITRE (1.78 PINTS) OR APPROX 1 kg

Piglets 3-6 weeks	4 mins	250 ml/min
Weaners/stores to 30 kg	2 mins	500 ml/min
Growers 30-50 kg	1½ mins	660 ml/min
Finishers 50 kg +	80 seconds	750 ml/min
Sows and boars	1 min	1000 ml/min

(one imperial gallon = 4.54 litres is 4,540 ml)

In hot weather, for sows and boars, double this flow-rate may be needed. These are best estimates, there has been little work done on ideal flow-rates, but these levels should not overtax the pigs.

Pigs which drink insufficiently will eat less food.

Check your flow-rates with watch and plastic squash bottles. Flow rates are too often too little or sometimes too much (1 litre per pig per day wastage costs up to £2 per pig to remove by slaughter weight!)

Watering the piglet up to weaning

•   **5 hours to 48 hours old**. Nipple displacement or bite valve supply. A flat dish is fixed to the slats in the farrowing pen, or on to a heavy iron plate on which the piglet stands to drink; heavy enough not to be nose-nudged around. This must be kept clean and water replenished several times a day.

Spend more time with the newly farrowed pigs!

The pen should *also* have either a sideways displacement (stalk) valve drinker, or better, a shallow Japanese style cup or 'leaf' drinker with a > 0.2 litre/min flow rate.

•   **36 hours old to 7 kg (weaning)**. The flat dish device is removed and the fixed farrowing pen piglet drinker used for consumption with a 0.3 l/min flow rate.

Table 4. PROVIDING CLEAN, REPLENISHED WATER IN A DISH FROM 6 HOURS OLD COMPARED TO CONVENTIONAL BABY PIGLET DRINKERS ONLY DURING THE FIRST 2½ DAYS OF LIFE

	Weaning wt at 24 days (kg)	ADG (g)	ADG (g) 24d – 25 kg
With dish and fixed piglet drinkers	6.28	205	452
With fixed piglet drinkers only	6.13	200	413

Source: Clients records 1999

Comment:

Since this trial was done, the 'Oasis' dome turkey drinker concept has been successfully tried on weaners (see page 144) but not on baby piglets. It would be interesting to see if a 'mini dome' suspended chicken drinker working on the same principle would give a similar benefit – and avoid the need for frequent water replacement in the "neonates' dish" concept – its main drawback. Perhaps water researchers might like to look at this?

Table 5. SUGGESTED FLOW RATES FOR OLDER PIGS

In my experience these should be sufficient and with the correct design of drinker, should waste the least water.

	litres/minute		
Weaner 7 – 25 kg	1.0		
Grower 25 – 50 kg	1.4		
Finisher 50 – 120 kg	*Bowl*	*Bite or Nipple*	
	1.9	1.7	
Dry Sow	*Bowl*	*Bite or Nose Press**	
	2.2	2.0	
Lactating Sow	*Bowl*	*6-10 litre trough+*	*Bite or Nose Press**
	2.4	1.8	2.0

Note: *Nose Press drinkers are not recommended as they can cause soreness.
+ The ample capacity trough in a farrowing pen is best, if designed correctly and set at the right height.

Number of drinkers per group

Larger groups of pigs are being housed together these days, so numbers of drinkers per group is becoming more important. My advice after observations on about 50 such farms is given in Table 6. A client did a trial after visiting a US unit where he saw a battery of bite drinkers in a large grow-out slatted-floor piggery containing 30 pigs per pen and the American owner felt they were beneficial at a late stage stocking density of 8 ft^2 (0.74 m^2) per pig. As the water lines needed renovation in two of my client's houses back home we did a trial which leads me to suggest the following (*Table 6*).

Table 6. SUGGESTED NUMBERS OF DRINKERS FOR DIFFERENT SIZED GROUPS OF PIGS

Group size	Nipple & bite drinkers	Bowl drinkers	Trough drinkers (mm)
Grower/finishers			
10	2	2	1 x 300
30	2	2	2 x 300
50	4*	4	3 x 300
> 50-200	6*	4 to 8	3 x 1 metre*
Per 15 sows	Not recommended (wastage)	2	300 mm per 15 sows

*These drinkers/troughs should not be sited close together but spaced some way apart in the drinking/dunging areas.

Water in CWF (Computerised Wet Feeding) systems

Fresh water should always be provided. The liquid in a wet feed 'soup' (even if fresh water rather than skim milk or whey) should be regarded as purely a vehicle to transport the dry feed, and the pigs need a separate water supply as would be normal.

As to water:meal ratio in the pipeline mix, Dr Brooks has done valuable work on this in the 1980s (*Table 7*) which also shows that the thicker the mix, the more the pigs require supplementary water from the drinker supply.

Table 7. VOLUNTARY WATER USE AND PERFORMANCE OF PIGS OFFERED LIQUID DIETS AT DIFFERENT WATER:MEAL RATIOS

Water:meal ratio	2:1	2.5:1*	3:1*	3.5:1
Meal intake (kg/day)	1.48	1.49	1.46	1.47'
Water from drinkers (kg/day)	1.26	0.78	0.44	0.24
ADG (g)	730	740	750	770
FCR	2.01	2.00	1.95	1.90
Total water drunk, ie from liquid feed and from drinkers (kg/day)	4.23	4.51	4.86	5.40

Source: Brooks & Carpenter (1986)
*These are typical water:meal ratios in pipeline feeding *ie* 2.5 W:M ratio = 4 lb meal/gallon (1 kg meal to 2.5 litres water) – which is pretty thick, while 3½ lb/gallon (1 kg meal to 2.86 litres water) is 2.9 W:M ratio, a more normal one and easier-to-pump over a distance. Some pipeline feeders with a long or tortuous circuit, pump at 3 lbs/gallon (1 kg meal to 3.33 litres water).

Some practical tips on watering which have stood the test of time over 40 years advisory work

- Make sure the near-to-farrowing sow is well-watered prior to farrowing. Put 4 litres into her trough twice a day from F–2 days to F+2 days.

- *Balantidium coli*, a relatively harmless organism where gut inflammation is concerned, is common in unsanitised water lines, but nevertheless seems to affect food conversion, ie this has often improved when it is removed by pressure washing drinkers/improved sanitising of the water lines/header tanks.

- So check the underside of leaf drinkers for slime. This also affects performance, especially at the end of the nursery stage (20-25 kg). Pressure washing removes it.

- Header tanks are a potent source of waterborne organisms, and must be kept clean and the water 'sweet'.

- Check the flow rate from header tanks. If too great, install break tanks; if too low, increase the outlet orifice size, *not* raise the height of the tanks (much less efficient).

- When examining nipple/displacement drinkers - look at the quality of the engineering with a magnifier. The best engineered will give *easily* the most reliable service.

- Measuring total solids in the water supply is a good watchdog and a quick and inexpensive test for TS can be done at a chemical laboratory. Danger warning is > 3500 ppm, signalling the need for sanitation and a recheck afterwards. If still high, filtration may be advisable in certain rural areas. Seek advice.

- I live in a hard water area. Farms around here get pipe 'furring' due to flow restriction from calcium and magnesium deposits, harmless to pigs but causes a gradual, often not noticed, fall off in flow rate. If this occurs, see (free) advice from the local water authority as it can be mitigated/ remedied.

- Every 2 years get a water *E. coli* test done. Target level is zero. If higher than this, sewage may be getting in to your after-mains line somewhere – usually a faulty or corroded joint – and is likely to cause an increase in disease. A real headache to sort out, too. Start with your veterinarian to get a test done and he can take it from there by calling in specialist advice, usually through the water authority.

WATERING THAT SOW IN LACTATION

This article caused quite a stir. I got more 'fanmail' (and a few brickbats from the drinker manufacturers!) in the few weeks after publication than for several years. The letters I got from lady pig-keepers were fulsome! That was nice!

Isn't it crazy the way we water sows in the farrowing crate! You think it isn't? Drinkers, providing you achieve a flow rate of up to 2 litres/minute are quite OK – so the textbooks seem to say!

I'd like to down-rate the current system we all use and ban the bite or nose-press drinker from the farrowing crate. My experience is that a really good big trough (not a small bowl) is preferable in several ways. Let me explain.

Fact: Sows should be encouraged, given a good growing litter, to eat as much as possible in lactation once they have recovered from the effects of parturition. (*Table 1*)

Fact: Sows are leaner these days. Selecting for lean has also selected against appetite. Appetite varies between genotypes. (*Table 2*)

Fact: Sows are rapidly becoming much more productive, mainly due to genetics (Chinese genes, etc.) but also due to improved nutrition and management of ovulation, fertilization and implantation.

Fact: Sows eat less when conditions are hot. (*Table 3*)

Milk needs a certain amount of water to form it – probably around double her milk yield per day. (*Table 4*)

Now think back to that sow. She's had an exhausting time – lost a lot of fluids giving birth. She's thirsty, and above all – *sore*! She *hurts*! Ask any woman who has had a baby.

It hurts to get up and down. She has to get up to feed, drink, defaecate and urinate, but she wants to get these over as quickly as possible and settle down again to rest, grunt-up and nurse her new family.

First mistake: We provide a drinker; which even at 2 litres/minute, a 30 litre/day demand takes over 25 minutes actual drinking time. No, **not** 15 minutes (30 litres ÷ 2 minutes) – nearly double, because she wastes a lot. Just watch them if you don't believe me.

Second mistake: Accessibility to the drinker is often ridiculously awkward. Try drinking a can of Coke at the level of your knees! If your drinkers aren't set just above the back level of any pig then she'll take even longer to drink her fill and waste much more. (Table 5)

She gives up!

This makes the sow very frustrated, you can almost see her saying 'the hell with it'. She sits down with relief but without taking in her desired fill. She doesn't drink enough, so she eats less, especially at temperatures over 20°-21°C, so milk yield may be affected and weaning weights can suffer.

The solution

A 6 to 8 litre trough next to the feed bowl itself holding 6 kg of food. The two are separate. Not a small leaf drinker bowl but a proper water trough. She can put her mouth in and just suck up 3-4 litres in a few seconds – just like an elephant. Then relax, her thirst slaked. A good example is shown on page 148.

Proof?

We took out the bite drinkers in a farrowing house, one side only, and replaced them with troughs (with an overflow hole to the passage-way). Compared to the bite drinkers across the feedway they drank more, wasted less, ate more (at 24°C;) and the piglets grew faster. (Table 6)

   If you feed dry, you *must* use water troughs, not drinkers. But better, feed wet by pipeline. Opinions vary if you still then need drinkers as well – I think in this case these are permissible, but set up and to the side of the wetfeed trough to deter overflow. And what about nose-push drinkers in a trough? Sure, these are better than bite drinkers as the Germans have found, but I am worried about the likelihood of sore snouts from these devices.

Table 1. FEED REQUIREMENTS 150 kg SOW WITH 10 PIGLETS GROWING 160-280g/DAY

	Week 1	*Week 2*	*Week 3*
Piglet Weight (kg)	2.5	4.0	6.0
Piglet growth/day (g)	160	214	285
Sows feed/day (at 19°C) kg*	5.1	6.6	8.0

*Ambient air temperature

Table 2. APPETITE OF 4 MAJOR EUROPEAN AND AMERICAN BREEDS (PIGS 25-100 kg : 55-221 lbs)

	Kg/Day
Breed A	2.84
Breed B	2.78
Breed C	2.79
Breed D	2.51

I can find no interbreed figures for lactation feed intake (they probably don't exist) but it is reasonable to extrapolate such figures to the dam's appetite, i.e. some genotypes will eat 13 to 15% less – at least. In hot conditions this is nearer to 20%, maybe.

Table 3. FEED INTAKE OF ONE GENOTYPE OF SOW AT 150 kg IN LACTATION (kg)

	Week 1	*Week 2*	*Week 3*	
Feed/day at 19°C	5.1	6.6	8.0	
Feed/Day at 26°C	3.9	5.3	6.6	
Deficit day (kg)	1.2	1.3	1.4	
Weight loss/week (kg)	4.4	4.8	5.2	**Total 14.4 kg**

This suggests, of course, that both feeds were not nutritionally-dense enough. That said, look at how appetitie fell when the temperature rose.

Table 4. ESTIMATED WATER NEEDS OF LACTATING SOWS TO SUSTAIN SUFFICIENT MILK YIELD TO SUPPORT PIGLET GROWTH (BASED ON 4.0 TO 4.5g MILK FOR EACH g PIGLET GAIN)

Scenario	*Milk needed/day litres*	*Water needed/day (minimal) litres*
5 day old litter of 8 piglets growing at 150 g/day	5.0	10
20 day old litter of 13 piglets growing at 345g/day	20	40
Higher levels seem to be needed in hot weather		

Table 5. WASTE OF WATER FROM THE SAME MAKE OF DRINKERS SET HORIZONALLY TO ONE SIDE AT 350 mm FROM THE FLOOR AND AN ADJUSTABLE DRINKER SET AT 15° AND 100 mm ABOVE THE SOW'S BACK-LINE

	Water used/day	*Lactation Feed eaten 0-21 days (kg)*
Low-Set	27.7 (+4.2%)	122
Correctly-Set	19.5	136 (+11.5%)

Source: Clients' records

Table 6. COMPARISON OF CONSUMPTION/PERFORMANCE OF LACTATING SOWS WHICH WERE TROUGH-WATERED V. DRINKER-WATERED, SUMMER CONDITIONS, FIRST 21 DAYS OF LACTATION

	Trough	*Drinker*
Water used/day, litres	23	18
Water spillage, litres	not measurable i.e. very low	3.6
Food eaten to 21 days, kg	139	127
Piglet weight at weaning, kg	6.03	5.81
Piglet growth rate/day (g)	223	212

Source: Clients' records

STOMACH TUBING

Stomach-tubing piglets has rapidly caught on in Europe.

Stomach-tubing (passing small quantities of nutrients directly into the neonate's stomach) has always been considered a very difficult, delicate and therefore rather dangerous operation, only to be used as a last resort if a dehydrated or nearly starved piglet needed to be saved and only to be done if you had plenty of time. We have woken up with a start to several important things:

- **It is not difficult.** One needs some self-confidence and the right equipment. And it is certainly delicate, but no more so than clipping teeth and less so than castrating.

- **It is not dangerous to the piglet.** I know stockmen who have successfully stomach-tubed over 500 of them. Some can remember one or perhaps two not making it, but those may have died anyway because they were too weak or from shock, certainly not clumsiness. I've done about 200 piglets myself and it isn't as difficult as it looks.

- **It is not time-consuming.** Collecting the colostrum from the sow takes a little time (about two minutes per 20 cc dose), but stomach tubing itself only takes maybe 1½ minutes per dose.

- **Far too many pigs die in the first 12 hours of life.** In fact, as Table 1 shows, just over half a piglet/litter dies by 12 hours and just under another half-piglet 12-24 hours after birth.

Table 1. BEST ESTIMATES (FROM STOCKMEN'S RECORDS) OF THE PERCENTAGE OF WEANING LOSSES FROM BORN ALIVES IN 14 MAJOR PIG KEEPING NATIONS, EXPRESSED AS TIME OF DEATH

First 12 hours	First 24 hours	2 days	3-7 days	7-14 days to weaning	14 days to	
37%	32%	12%	8%	6%	5%	
Expressed as piglets on a litter size born-alive of 10.6						Total
0.53	0.46	0.17	0.11	0.09	0.07	1.43
——— 1.16 piglets ———→			——— 0.27 piglets ———→			

Most pre-weaning deaths are early in life

So from these figures 81% of all your mortalities to weaning could well die within the first 24 hours. Even if you reduced this neonatal loss by about half a piglet you would bring down a 13.5% mortality-to-weaning figure to 7.26%! But I prefer to look at losses in absolute terms. On a 10.6 born-alive average, 13.5% mortality to weaning is 1.43 piglets/litter lost by weaning, while 7.26% is only 0.78 piglets lost or 143 more piglets saved per 100 sows/year! A big, big benefit. In the UK, in cost of man-hours vs. extra margin obtained, it is a 7:1 cost:benefit ratio. It pays hands-down to be there at farrowing!

Some piglets need stomach-tubing because:

- They don't get a sufficient dose of colostrum soon enough.

- They don't get the 600 kcal DE intake of energy needed to sustain muscle vigour in the critical and competitive first 24 hours.

- Their body temperature quickly drops to under 98°F (from 102°F normal), immediately putting them at risk from hypothermia despite warming — it is internal warming they need, even more than outside heat.

Note: The flotation collar device and warm water immersion at 110°F for 2-3 minutes is an excellent idea for those too far gone and flaccid to take stomach tubing, which can be done once they are warmed up and dried off.

Realisation

We have now realized that 10-20% of all piglets born (6-12 piglets in each farrowing batch of, say, six sows) can be at risk. These are the very small ones which are usually born later or last; those who find it difficult to secure a teat and those born in the small hours of the morning. By the time the stockman arrives at 7 am, the piglet's gut apertures could already be "closing" to the large molecule size of colostral antibodies. By midday, the door is shut to most of them. They are immune-deficient at a critical stage of survival.

Stomach-tubing such piglets allows them to catch up with their stronger littermates; to acquire sufficient colostral antibodies and extra fuel to invigorate them to withstand competition at the udder better. Small piglets can, as you know, be quite vigorous, providing they have the muscle power to do it. A full belly of colostral milk turns weak pigs into competitive little animals within an hour.

Stomach-tubing can also be helpful to older piglets that are falling way behind. In this case, ordinary sow's milk or a "colostrum substitute" based on cows' milk is used.

Technique

1. One first has to **collect the colostrum** during the short period when letdown is continuous. This is during and very shortly (10 minutes) after farrowing. Collect about 80-100 ml into a wide-necked container from four to six teats.

Squeeze the teats gently and never milk-out a cistern, however free the milk flow, or a piglet may be deprived of its normal feed. Clean hands!

2. **Store the colostrum** in plastic AI squeeze bottles in a refrigerator for 48 hours or in a freezer indefinitely. Freeze quickly and *thaw slowly* using a warm water bath — never microwave colostrum to bring it up to body temperature again (102.9°F) or the globular immune structure is damaged. The colostrum can be from any sow in the herd, *but interherd exchange is not advised*, and on the very biggest farms, intersection exchange too.

3. **Prepare the equipment**, which must be clean and dry before use. With your hands and clothes clean, use a plastic 20-ml syringe with 25 cm (10") – no more, no less – of narrow-gauge silicon plastic tube (model aircraft fuel line is excellent). Lubricate the tube with corn oil or olive oil. Warm the bottles of colostrum to body temperature. Fit the tube snugly onto the syringe.

4. **Check that the piglet selected hasn't got a full stomach** or you will drown it! (Get to know what an empty and full stomach feels like before and after suckling.) If it *has* eaten, put it in a warm box for 1½ hours.

5. **Introduce the tube.** Place it on the end of the piglet's tongue and guide it gently down the pig's throat. The piglet will swallow the tube – just guide it down. At 10-12 cm (4-5"), you will feel it pass the sphincter valve at the top of the stomach. Do not pass more than 15 cm (6") down, leaving 10 cm (3-4") outside the mouth.

6. Tip the syringe up and **pour 20 cc of colostrum in,** then steadily push the plunger down for two to three seconds, and withdraw the tube. Take care not to introduce air into the stomach.

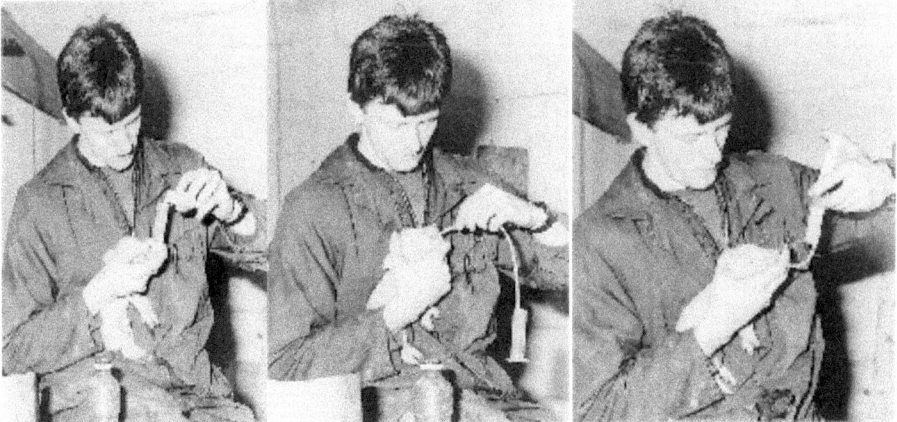

Stomach Tubing (1)
Gently insert the silicon plastic tube over the piglet's tongue – the piglet will swallow it – just guide it down. At 12 cm (4-5") you will feel it pass the valve at the entrance of the stomach – allow 1"-2" more and stop.

Stomach Tubing (2)
Holding the tube in place, up end the plastic syringe and pour in 20cc of previously-collected colostrum. I prefer to support the loaded syringe on my forearm or if the tube is long enough, over my shoulder.

Stomach Tubing (3)
Gently press the plunger for a few seconds. Take care not to push air into the stomach. Gently withdraw the tube.

7.   If things go wrong (a common fault is not passing the sphincter before dosing), and milk appears at the mouth, quickly withdraw the tube, face the piglet downwards so that excess milk runs out and gently compress the lungs in case any was breathed in. Examine the piglet. Leave for a while if in distress.

Stomach-tubing is much more effective than "drenching" by mouth. Much colostrum is wasted with drenching and it is easy to get liquid into the lungs. Surprisingly, stomach-tubing needs less skill to achieve the same result and is quicker.

Very weak piglets should be stomach-tubed three times at hourly intervals. Other small piglets should be dosed depending on how much milk they can obtain naturally.

About 10-15% of piglets needing stomach-tubing will be too weak (usually too cold) to withstand it. Use the flotation collar and warm water treatment to save these otherwise healthy but moribund individuals.

DOUBLE-TAPING SPLAYS PAYS OFF!

Splay-leg is on the increase in Europe. Target level incidence is less than 2% (in 0.5 to 1% of farrowings) but this threshold level is being exceeded on most of the farms I visit. Splay legs are a real nuisance and turn-off to stockpersons.

Reasons for splays

(a)   Genetic

The reasons for this increase today is not hard to find, as splay-leg is positively correlated to meatiness in male lines – as our lean yield improves, so the number of splay-leg newborns also rises. Occasionally one can identify a rogue boar from his records, but not usually before 6-7 months use. This must mean that splay-leg could be an increasing problem, as it is going to be a long and expensive job to breed it out of seedstock male lines.

(b)   Nutrition

Splays also are higher when foods are high in polyunsaturated fats (P.U.F.A.) and/or low in vitamin E. Protection must occur both in the lactation diet and in pregnancy, so ensure that all sow feeds have 30 mg/kg vitamin E added for each 1% PUFA in the diet, and this should be a minimal level in all creep/prestarter feeds too. In some countries farmers won't know the likely PUFA content of their feeds, so a rough safety rule is to check that at least 5 IU of vitamin E (i.e. 5 mg) is present for each 1% declared oil above 3%, ie declared oil 5%; 5 x 5 = 25 IU. These vitamin E levels need to be *added* per kg feed of 14% moisture. In cases of less-than-fresh food these levels may need to be substantially higher.

Secondly; producers should check with their nutritionist that any other vitamins thought to be associated with the sow's nervous system are at adequate levels, especially if the food is damp, and ask themselves if moulds could be associated in any way. Mycotoxins (mould poisons) are thought to be involved in some cases of splay-leg, so mould inhibitors should be present in all sow foods.

Third; in some less well developed countries sow foods low in energy are responsible. This is rare in Europe, however, unless the harvest has been bad and wet.

Fourth; the effect of stress on sows in pregnancy, particularly cold, bullying or deprivation of comfort, food or water can cause an increase in splay-leg which disappears as soon as the conditions are remedied. This is almost certainly a nutritional 'blocking' interreactive effect.

(c)   Disease

Strangely, parvovirus vaccination has lowered it. The reason is not known.

(d)   Environment

Slippery floors and all-metal slats make splay-leg worse. Use mats, old carpet squares or heated pads for a few days after birth at least, or longer if possible.

Treatment

So we are all going to get splay-leg on our farms – some more than others. Thus prompt and correct treatment is important and the writer has made a special study of this in the UK.

1.5% of our born-alives are starved out or crushed even though their hind legs are taped. That's too big a waste – about 30 pigs lost on a 100-sow farm each year.

In the past few years I've been working with farrowing house attendants on four farms. We have cut the loss by 90%!

Out of 16,500 born-alives, we had 322 splays – and we lost only 17! How did we do this? To understand the process you must realize there are 4 separate techniques you can use.

1)   Single-taping

There are two problems with taping (as distinct from double-taping which I'll describe shortly). First, having taped the piglet's back legs with ordinary surgical tape or a 'Velcro' strip just above the pastern joint so that the legs are 4-6 cm apart, the stockman tends to plug it on a teat and go off on his rounds.

Not good enough! In fact, this is only half the job, as that piglet is now too stressed and too immobile to compete.

So he loses his vital 40cc of early colostrum. In a *correct* single-taping procedure, we dry the piglet off, tape it up and then milk about 20 cc of colostrum into a clean, dry syringe, or have it available in a colostrum bank in the fridge – colostrum from any sow on that farm or farm section will do – warmed up *slowly* first, of course! We dose the taped pig with that colostrum. See how to do this in the Stomach Tubing Section, pages 159-162). And the dose needs to be repeated an hour later (within 1-1½ hours of birth, if possible). Problems do occur with overnight farrowings, but even then, it should be the first job on arrival in the farrowing room.

But even with the colostral feeding, we are not finished. Next, all treated splay-legs are moved to a milky sow and allowed one full suckle while her litter is shut in the creep area. Each splay may need assistance to "settle" at a teat.

Lastly, the pigs are returned to their own litters and the tape is removed in two or three days, or earlier if locomotion is seen to be activated.

You will lose very few splays if this routine is followed. Just taping and hoping everything will be OK is not satisfactory, I'm afraid.

2) Massage

This practice has been recommended by Philip Blackburn, an experienced British pig vet. The technique is explained in Table 1.

I've done quite a few splays this way and it certainly works, especially in the really bad cases which have front-leg trauma as well. But it takes time, perhaps eight minutes of massage in three or four two-minute sessions. However, it is a humane and positive technique compared to the cruder taping, which must stress them considerably. I recommend it for the real "stiffs" and for the farmer lucky enough to only come across the occasional splay.

3) Double taping

This is the technique which gave us the negligible losses from splay-leg on 4 farms over a year's period. The big advantage is that the period of restriction is cut to a few hours – half a day at the very most.

Table 1. THE BLACKBURN MASSAGE TECHNIQUE

1. Mark the piglet (you'll need to pick him out again).

2. Sit or kneel down and place the piglet, head away from you, with his chest resting on your knee, which is inclined downwards.

3. Hold the piglet with a hind leg in each hand, gripping the lower legs with your last three fingers, and the crutch with your index finger.

4. Vigorously massage the lumbar muscles with your thumbs.

5. Drop down either side of the tail to the ham muscles, dealing with the muscles over the pelvis as you go past it and using your index finger to massage the inside of the leg.

6. After about two minutes, the muscles will relax and each leg can be flexed quite easily. What you have done is improve blood flow to the muscles and activate nerve responses.

7. If the forelegs are affected, hold the upright piglet between your knees and give similar treatment to the front legs.

8. Encourage the piglet to stand up and proceed with the colostrum-dosing as per normal.

9. This may need to be repeated three or four times during the first day.

In double taping, the rear legs are taped normally, and then, usually with a little help from an assistant to hold the piglet while you do the taping, the legs are gently moved forward under the piglet's trunk and slightly to one side (see diagram). Then a second band of tape is placed over the body, just forward of the pelvis, so as to hold the legs close to the body. The piglet is now well and truly trussed!

This means you put it into a heated straw box, give it its dose of colostrum and leave it there, with any similar unfortunates, for three to four hours. This is important, because it cannot move far and will certainly get crushed if left with the sow.

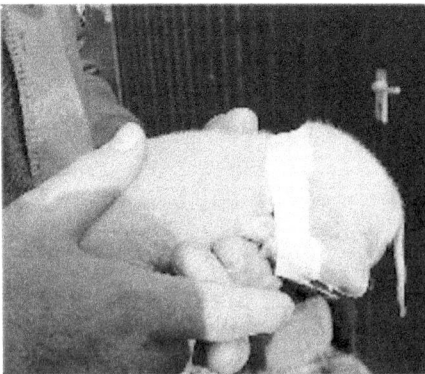

A double-taped piglet. Both the hind legs are taped together, and then taped to one side across the back.

After only 2 hours in the hay box and on removing the tapes, this piglet totters away, cured.

Why immobilize it? Strange, but it seems that attempts by the piglet to use its bound rear quarters stimulates the affected nerve endings – it seems to feel no pain so doing – and thus regains control quite quickly.

4)   The exercise chute

Another useful idea I have used, and which I saw in Denmark, is to make a wooden exercise chute, with two vertical boards nailed to a wooden base, about the width and length of a newborn. Holes are bored through the back of the side boards on which the upper part of the piglet's back limbs would rest, and a piece of stiff rubber tube inserted. This results in both back legs being suspended off the ground rearwards, the piglet resting on its chest and front legs. A second set of holes are bored to take a second rubber rod across the small of the piglet's back to hold it face down and prevent forward movement.

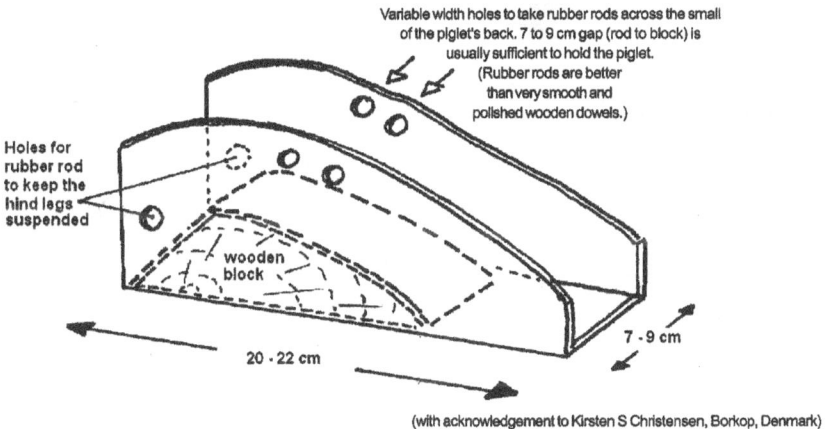

Variable width holes to take rubber rods across the small of the piglet's back. 7 to 9 cm gap (rod to block) is usually sufficient to hold the piglet. (Rubber rods are better than very smooth and polished wooden dowels.)

Holes for rubber rod to keep the hind legs suspended

wooden block

20 - 22 cm

7 - 9 cm

(with acknowledgement to Kirsten S Christensen, Borkop, Denmark)

A home-made exercise chute.

The piglet struggles to get free but the rear trotters are above the ground, or lightly resting on it, and providing this is a smooth surface, it cannot get free, but works its hind legs vigorously.

Some practical advice goes with the device.

•   Piglets should be over 900g in weight.

•   Place it in the creep area, which encourages the piglet to get free. Otherwise it may just go to sleep and give up trying to do its 'ten minute workout'!

- •   Check the condition of the piglets after a minute or two.

- •   Do not use the device on moribund, very weak or chilled piglets (warm these latter up in a 'bath-and-collar' pail).

About 10-15 minutes on the device will cure many splay legs.

A suggestion for a home-made excercise chute is given on the previous page.

STIMULATING GILTS

I'm in two minds about adding this subject to the series as, in all fairness, the subject is dealt with in depth in many textbooks. Nevertheless, when I read the dozens of pages which have covered the area of getting gilts mated in prime condition, I'm often left with a feeling of unease that some management factors I've felt to be important in my own experience have been left out, or not emphasised enough. So perhaps a few thoughts on this might be useful.

Casual?

Are we getting a bit casual these busy days? The textbooks tend to advise fence-line contact between gilt and boar, running a highly sexually active boar past the group of pre-pubertal gilts or penning him alongside them. I've always found that closer contact than that is worthwhile, and that the faster the gilts have been grown the more important this is.

For gilts 'in with', not 'along with'

Closer contact? By this I mean putting the boar *in with* the gilts, under supervision, of course. 'Sex scents' (boar pheromones) are frequently made much of by the textbooks. Okay, fair enough, but the boar's *saliva* contains them in abundance and the gilts should be able to have time to physically explore the froth too. Research by Hemsworth & Hughes in Australia suggests that 20 minutes a day physical contact is much better than (if it is done at all) the usual 5 minutes. As with so many things in pig husbandry, putting enough *time* by to do things properly is vital.

This in-pen socialising becomes more important the larger the group of gilts to be stimulated. Today – in Europe anyway – is the day of the group, and while I've always said that six gilts together is enough, with straw becoming more commonplace in breeding sow housing, I now see tens and the very occasional twelve. There again, ten gilts milling around whom they see as Brad Pitt for 20 minutes can be a bit of a circus – it is much easier with six! As Brad Pitt, no doubt, will tell you!

Twice in hot weather?

Summers are also getting hotter (global warming?) in Europe, and this dulls anyone's senses! So it is with pigs. During this hot weather, and if the morning rush is too busy to supervise the 20 minute chat-up, try exposing the boar twice a day, after the morning shift and before clocking off in the evening. For me this seems to have solved some weak-heat problems in gilts.

Next, the boar. The textbooks rightly state that the one chosen to stimulate gilts should be full of libido, such as a herd sire 'favourite' with the stockman which is regularly used (fully-AI units often have a problem here). In any case it is wise to delay the use of such a boar for gilt stimulation purposes until he is 9-10 months old, even though he may be technically 'mature' at 6 months, as the textbooks advise. Too soon! He needs the practice of sexual arousal to *release* the pheromones, not just produce them.

Habituation

The (older?) textbooks advise penning as libidinous a boar as possible alongside pre-pubertal gilts. But is this wise? Even if you change him periodically, the females get used (habituated) to the signals. It is thought that the sexual signals of sound and smell do not in fact travel far, so the boar's pen must not be too far away, but if he is kept two or three pens away and then brought *in* to the group of gilts for 20 minutes, you get the best of both worlds.

Weaned sows are different. They have not had time to get used to the boar's presence during the short 3-6 day interval between weaning and mating. Having the boar alongside these ladies is permissible, I guess.

Air quality

I favour an atmosphere where pheromones can pass freely inter alia, but not become stuffy and overloaded with ammonia and H_2S gases in winter, or with a gale of cooling fresh air blowing through in the summer, a common failing in the tropics. The answers to both would be to keep the gilts' pens cleaner in cold weather, and place the cooling air through a polythene trunk in summer so that a full 'tunnel-ventilation' effect in very hot conditions is not needed, especially if droplet water spraying is installed.

If the air quality is difficult to control, then finding time to practise the physical boar presence *in* the pens becomes even more worthwhile.

HEAT DETECTION

30 or more years ago most textbooks on pigs were written by farmers and were full of practical tips but weak on science. Now most weighty tomes on pigs are written by scientists, many of whom understandably have never farmed pigs of their own, so the pendulum has swung the other way.

This book tries to redress the balance a bit, so here are some practical tips on the vital task of heat detection.

Things are rather different with gilts, however, which are new to the whole process. Read Stimulating Gilts on pages 169-170.

"EVER SERVE YOU RIGHT !. . . " . . . said the notice I saw pinned up by the room service staff in a French hotel! Where sows are concerned, it does serve you right if you don't serve them right!

Heat detection is the key. Knowing *exactly* when is the start of oestrus. Here is a reminder of 12 practical points which will improve your success rate.

A heat detection checklist

1.   Use a boar to help you – he is better at it than you are !

2.   Boar to the *front* of the sow, not the rear. He likes the smell of the rear, but *she* likes his face to the front !!

3.   Start detection early. As soon as gilts are delivered; from 3 days after weaning in sows; then detect both twice a day.

4.   Best housing arrangement is to place rebreeding sows[1] one metre away from boars, in full view and without pheromones being blown away by over-fast ventilation or masked by faecal/urine smells if the ventilation is shut down in cold weather. You could use a product called Deodorase (Alltech) if so, this has been proved to help raise conception rate. Added to the food it reduces ammonia from the faeces by a third. (It takes about 3 weeks to become fully apparent.)

---

[1] There is different advice for *gilts*. (See 'Stimulating Gilts', pages 169-170).

5.   Detection is easier if the sows have no sight, sound, touch, etc, of boars one hour prior to checking. Not usually practicable but a must with 'difficult' sows. Try it!

6.   As long as females in groups are not obviously on heat, allow the boar into the pen, otherwise he will serve one, usually too soon.

7.   Groups of sows need 1.9m^2 (gilts) and 2.8m^2 (sows) fleeing space to avoid proximity stress which dulls the effect of oestrus. Try not to mix differently-sized females – this causes stress, too. Stress hormones tend to act against sex hormones.

8.   I find it is very important to have bright light (350 lux) shining into the sow's eyes for two thirds of the 24 hour day length.   Four out of five breeding sections are badly under-lit (100-150 lux.) 350 lux is about as bright as a 4 ft 100w fluorescent tube light in a smallish kitchen – pretty bright. Many breeding sheds/sections are nowhere this brightness. Use an (old fashioned) photographer's light-meter as I do, calibrated by using the 'kitchen' example above – holding the light meter about 3 ft below the tube, and using this reading as your guide when held just above the sow's eyes.

EYES, EARS AND TOUCH

9.   *Use your ears* to detect a 'chattering' (calling) female.

10.   *Use your eyes* to detect restlessness. Pre-oestrus occurs 12-24 hrs before oestrus proper. At pre-estrus the gilt's vulva almost always **swells**; but maybe or maybe not with sows. Most vulvas **redden** however; nervousness commences; riding or being ridden may be present. There can be *clear* discharge from the vulva.

11.   *Use touch* to test for back pressure, and part the vulva to check for discharge inside the labia. Clean hands, please.

12.   Be quiet and patient. Take *time* to detect heat properly, and you'll hit ideal service smack on.

And a thirteenth (I find unlucky) idea is to use chemical aids. Don't do it, get expert at the natural way!

The economics of good heat detection

Is at least a 5% improvement on your farrowing rate *even on good units*. This is because many stockpersons are not as good at heat detection as they could be, largely due to their not spending enough time with the pigs "because I've got more urgent things to do which won't wait." Top stockpeople agree that heat detection is one of the top three tasks (breeding and farrowing are the other two) and devote top priority to it, sorting out the "won't wait" problems to their satisfaction. Top section heads are good problem solvers and sort out the priorities, of which this is one.

So what's 5% more worth? Quite simple; it is 5% more piglets born per week, per month, or whatever *for the same amount of investment in food, housing, borrowings and overheads* (except labour).

Labour costs, *vis-à-vis* more time spent in heat detection, rise by 10% but labour cost is only 14% of total costs on most units. So for a 1.4% rise in total costs your increased income is at least 5%, an REO of over 3.5:1. In nett profit terms on my friend's unit a 5% lift in farrowing rate netted them 20% more profit, more like a 5:1 return where it counts most – money in the bank. And if you want to pay a quarter of that as a bonus – why not!

That's why heat detection is important!

Some straight talking on light

The advice in this book suggests modifications to current knowledge and provides some extra information which I think doesn't exist in some textbooks. The following observations are the only examples in this book where I am completely at odds with the views of some scientists; in this case about the effect of poor lighting on performance.

From experience on hundreds of farms I am quite convinced that those of them who say lighting intensity and duration is of little consequence are unequivocally wrong!

Here goes!

I am a firm believer, in sows, that the 16 hour on/8hour off diurnal lighting pattern is proved to be of value in practice; that a light intensity of at least 350 lux is needed during the period of post-weaning across the service period (and for stimulating gilts); and that 400 lux or more on the same 16/8 hour pattern in the farrowing house can improve litter weight at weaning (by up to 15%?)

Virtually on every occasion where I have persuaded the producer to beef up his lighting in these areas, and have then followed up the advice given, the feedback has been enthusiastically positive. *Everybody* agreed it definitely made a difference in sow productivity and easier conception. Weaner weight – maybe, not always.

Petchey in 1987 first noticed it (*Table 1*) and my own experience parallels it, with 15% - 20% more sows getting mated satisfactorily within that vital 5 days from weaning which so significantly reduces the herd empty day gap and so magnifies income over costs.

Table 1 INFLUENCE OF SUPPLEMENTARY LIGHT ON SOW REPRODUCTIVE PERFORMANCE

	Control	*Supplementary light*
No of sows weaned	164.0	163.0
Days to mating	5.9	5.5
Mated (0-5 days)	68.5 %	83.0 %
Mated (6-10 days)	26.8%	10.9%
Mated (> 10 days)	4.6%	6.0%

Source: Petchey (1987)

My suggestion for a lighting diagram to assist heat detection and facilitate conception is given below.

Lighting a mating / breeding house

100 watt fluorescent strip lights, white
Aim for 16 watts per m² (1.6 watts/ft²)

Place the lights so that the majority of the light falls via the eyes

Lighting pattern

On maximum 16 hours/day
Off for 8 hours/day

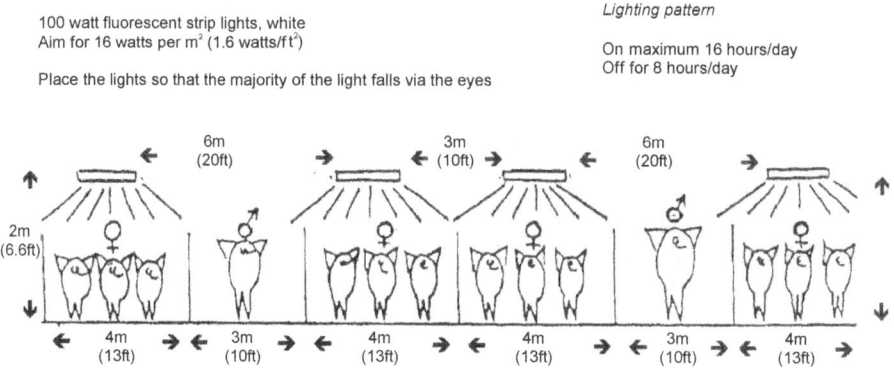

Practical advice.   For yarded groups, if the tubes have to be set above 2m to avoid damage by tractors, etc, double the wattage for each metre of height between two and up to 3.5 metres. Better, use easily altered chain fixings to save power/ overheads. Keep the tubes *clean*.

Reference

Petchey A M (1987) 'Supplementary Light and Pig Performance' pps 17-19; Farm Buildings Progress April 1987.

The escalating problem of disease

BUILDING A STRONGER IMMUNE BARRIER

Most pig farmers - in fact all livestock farmers - are too ignorant about the basics of the very complex subject of immunity. This is costing them millions in lost productivity as diseases strike home, especially those caused by viruses.

I wrote this article years ago. Sadly, it's message is just as relevant now as it was in 1988.

"Don't tell me the textbooks don't cover immunity!" you say, "Why, they are full of it these days."

That is true, and they do cover it very well indeed – but (in my opinion) in far too much detail. Sure, the whole subject of immune protection is extremely complicated, and immunologists (just like statisticians!) have not yet found a way of communicating their involved subjects in a way which the layman – in our case the farmer and his co-workers – can understand and put into practice.

Meticulously-correct scientific descriptions are all very well, but tend to defeat the objective if the producer cannot fathom it all out, not understand and thus fail to become convinced as to what he should do, and so doesn't manage to apply the worthy advice. That's where we are at on immunity just now.

Those of us who have spent a lifetime at the sharp end of pig production – through sheer experience of what seems to work and what doesn't in a wide variety of on-farm circumstances (with some of us also involved in widely different overseas climatic conditions) – now follow certain basic rules so as to let the pig build the correct immune barrier to the farm circumstances at the time, to ensure our clients get the message. Admittedly there are quite a few of them (see later) but they remain basic nevertheless.

Get your head round them and see disease levels fall!

Keep it simple

Despite immunology being one of the most intricate and involved scientific '-isms', it has been quite surprising to me that the core of the subject can still be effectively summarized to help the farmer in just a few sentences. 'Oversimplification', the scientist says. Well, maybe. 'Yes, but...' the veterinarian adds, pointing out from his own sharp-end experience when some aspects of these simple guidelines haven't worked, or *seemed* not to have worked. 'Sure,' I reply, 'It doesn't always happen especially if *only one of them is used solus*. But why not try as many of the suggestions as you can, then reassess.'

Viral diseases winning?

We are beset by diseases these days. We always have been of course, but today viruses seem to be outrunning us. Why? We front-end advisers are becoming increasingly convinced that pig producers are doing two major things wrong.

We are pushing the pigs too hard; stressing them too much. This is lowering their ability to fend off some particularly nasty organisms which are negotiating their way round the protective barriers we have created through experience and which served us reasonably well in the past.

As pig producers we haven't yet accorded sufficient priority to understanding immune status. If you don't understand why some practices are *essential* these days, you won't spend the necessary money to make it happen on your farm. Sorry, but I see this happening every week of my advisory life these days. It is still a big job to persuade people to do what is necessary. To spend what is necessary – in money and time.

It is significant isn't it, that after working in some 10 major pig industries world wide, I find *it is the ones who have suffered long bouts of low/ no profits who now have the worse disease headaches due to inadequate investment* caused by lack of cash flow. In most cases the deficient areas are **those that impinge on immune status.**

Immunity: The key points

So what are these, each in one sentence? They are not in order of importance as *any* of them can be N° 1 on different farms. Just do as many of them as you can. If you do, you will be surprised how much better your disease picture can become.

There are two distinct and different parts to the level of disease challenge:

(A)   Helping the *growing pig* by *lowering* the immune challenge;

(B)   Supporting the *breeding pig* by *strengthening* their immune defences in the following ways: -

- Reduce the need for immune stimulation from other pigs. Older pigs are a major source of challenge to younger pigs, so segregate by age.

- Adopt an all-in, all-out policy wherever possible. But do it properly!

- ***Thoroughly*** clean and disinfect weaner, grower and finisher accommodation between every batch. This includes fogging enclosed air spaces and sanitising the water system.

- Reduce dust levels in pig houses. Dust particles are pathogen 'taxis'. More dust, more risk of disease overcoming immune defences.

- Where continuous production has to be practised, institute short production breaks either by selling young pigs, or following the 'partial depopulation' idea. Utilise these breaks to clean and disinfect thoroughly.

- Adopt tight on-farm biosecurity. Most farms are still too lax, especially with visiting ***vehicles***!

- Avoid stressing the pigs. A huge subject. Study it. Do periodic 'stress audits'.

- Don't overcrowd / overstock. This particular cause of stress weakens the immune 'wall'.

- Have plenty of 2nd to 5th litter sows in the herd. These will have a good immune wall, naturally. So let nature help.

- Follow a new-stock induction programme agreed with your veterinarian. Even so....

- Don't push gilts too fast into first service, and therefore ...

- Set up gilt pools. They 'calm you down'/encourage good management.

- Be there at farrowing so as to practise immunity-encouraging techniques early-on.

- Don't wean too soon. In my opinion 21 days is exactly when we ***shouldn't*** wean! Everything is changing around this time in the piglets, endangering stress overload especially for the sow and her litter.

- Good nutrition is vital of course, but study how some of these 'new' additives can help the immune status directly, like organic zinc and, now possibly, several others, like chromium and correct fatty acid provision.

- Continually dialogue with your veterinarian, as immune defence patterns change with time. Nature never stands still, thank Goodness, otherwise you and I wouldn't be here!

To help you by making immunity less scary....

Some immunological terms

Humoral immunity = B cells, lymphocytes
Memory cells which remain behind after an infection, recognize the reappearance of the pathogen and quickly call up the correct defences.
Cellular immunity = T cells
These stand guard against pathogen challenge, are limited to body cells in various tissues susceptible to pathogen ingress.
Systemic or mucosal immunity
Local humoral or cellular antibodies ideally present when body surfaces are exposed to the outside – nose, throat, gut, outer reproductive tract.
Antibodies
Protein structures (IgA, IgM, etc) which fight antigens and unless overwhelmed, prevent disease
Antigens
Foreign material which triggers the body's defence mechanism – pathogens or vaccines.
Active immunity
After exposure to infection, stimulated antibodies remain in the sow which are transferred to the offspring via colostrum for a while in the form of antibodies IgA, IgG, IgE, IgM etc. The dam is *active* in passing on the immunity
Passive immunity
The piglets accept the antibodies (*i.e.* are *passive*) and this lasts as long as the maternal antibodies survive. As no memory cells (lymphocytes) are provided or formed so the immunity is not permanent.
Acquired Immunity
After a pig recovers from disease or vaccination it develops acquired immunity, triggered by antigens.
Phagocytes
Cells which ingest and so destroy pathogens.
Macrophages (white bloodcells)
Large immobile cells, usually originating in bone marrow, which become actively mobile when stimulated by inflammation, immune reactors and microbial products.
Cytokines
Messenger proteins which control macrophages and lymphocytes.

FEEDBACK AND IMMUNITY

If feedbacking is often frowned at these days, why write about it? Because it is still commonplace on many pig farms worldwide, and is often done erroneously.

Most textbooks are understandably shy about discussing feedback. 'Understandably' because, in my opinion…

- There are times when it is definitely inadvisable (see Table 1) and readers may be encouraged to follow one protocol when they should be doing another.

- Conversely, feedback can help considerably to recover from a TGE storm, or *neonatal* scour in piglets caused by viruses – although, not necessarily in the case of later preweaning scour.

- Portions of sow's afterbirth are often given to gilts by farmers. There is more hope than science here, I'm afraid, and where leptospirosis is concerned, now a growing disease worldwide, feedback is known to spread it.

Table 1. FEEDBACK OR NOT? CIRCUMSTANCES WHERE FEEDBACK OF SMALL AMOUNTS OF FAECES SCOUR, UNTREATED PIGLETS INTESTINES AND AFTERBIRTH ARE OR CAN BE DANGEROUS

➢ Swine dysentery	➢ Pyelonephritis
➢ Clostridial dysentery	➢ Eperythrozoonosis
➢ Erysipelas	➢ PRRS
➢ Leptospirosis	➢ Toxoplasmosis
➢ Metritis	➢ Vulva-biting in sows

Diseases which have been mitigated by feedback but only under veterinary supervision
- ➢ TGE
- ➢ Neonatal Scour
- ➢ Some other persistent enteric infections
- ➢ Cases of much more scouring in gilts' litters than in sows'

In fact John Carr, a well-known Australian pig adviser, wrote to me saying that most Australian veterinarians are now actively discouraging feedback completely. "Almost all the herds I am involved with in Australia and New Zealand have abandoned feedback altogether".

Feedback tends all over the world in my experience, to be done on a 'by rote' basis where one technique is practised unchanged for many months, and where the veterinarian is not consulted and used as an interpreter of the disease profile on that farm or in its immediate district. That is not using feedback (as and when needed) in a proper or effective way.

A part of the whole

Feedback is just one tool in the armoury of measures we can use to combat disease by strengthening the breeding herd's immunity. As Table 1 reveals *you must consult a veterinarian before practising feedback*. The disease pattern changes. Some diseases are better blocked by vaccination, others not. Your vet will have your current disease profile in mind, and can advise on what his tests reveal. This may or may not involve a degree of feedback under proper control.

"Used this way the vet costs too much!"

At this junction farmers protest about the cost of disease profiling from the vet in addition to his 'fire brigade' action and general advice. Because of this consumer-resistance, I recorded disease incidence and increased veterinary costs on three farms on a before and after basis (Gadd, 2003). While the cost of the veterinarian rose by 22%, his medication costs *fell* by 16.6% giving an average 58% more cost to the *vet/med* section of the balance sheet, which is what the farmer complains about. *However*, the estimated disease costs fell by 63% giving an overall improvement to the overall balance sheet of 51% on one farm, 9% on the second and 43% on the third. Good value!

One very interesting figure to appear was that the extra cost of planned preventive medication (including some feedback) was *lower* than the previous reactive curative medicine – by 9%. The full set of data is in my textbook 'Pig Production Problems' page 401, but I enclose the table again here as it is well worth study (*Table 2*).

Methods of feedback

This has been very clearly set out in 'Managing Pig Health...' by Muirhead and Alexander (5M Enterprises) 1997 pages 62 and 485, and the advice cannot be bettered. To repeat it here would be plagiarism – and this majestic work of reference should be on your shelves, anyway.

Table 2. BEFORE-AND-AFTER RESULTS FROM USING A PIG SPECIALIST VETERINARIAN TO DISEASE-PROFILE 3 FARMS, WITH EXTRA VACCINATION & RE-MODELLING EXPENSES COSTED IN. (US$ PER SOW)

Farm	*Before*			*After*		
	A	*B*	*C*	*A*	*B*	*C*
Estimated cost of disease per year*	284	186	300	80	96	109
Cost of veterinarian	8	3	12	30	27	31
Cost of vaccines & medication¹	26	18	30	18	20	21
Cost of remodelling (over 7 years)	–	–	–	27	45	33
Total Disease Costs (US$)	318	207	342	155	188	194
Difference (Improvement) %	–	–	–	51%	9%	43%

*Disease costs *estimated* from items like the effect of post weaning scour and check to growth on potential performance; respiratory disorders, ileitis, abortions, infectious fertility, etc.
¹ Note that the cost of planned preventive medication was *lower* than for reactive curative medicine.
Source: Clients' records and one veterinary practice

New techniques

As we get better at longer induction (acclimatisation) times, which allows the gilt to be grown less frenetically to the necessary heavier service weights, and using techniques like buying weaner gilts as well as building gilt pools – then the need for feedback would recede. And, I must add, understanding the whole complex process of immunity better, too, because many producers are nowhere near as proficient in this vital sector of profitable pig production as they are at the equally vital tasks of breeding and reproduction, for example.

Conclusions

So what am I saying about feedback?

• It has its place in the fight against disease.

• It must only be used under the supervision of a veterinarian.

- It will be used less in future as new techniques make it less necessary.

- If used, it is sensible not to adhere to just one method or system *in perpetuam*. Take advice on this.

- Farmers must get more proficient in understanding immunity, and how the various countermeasures, including feedback, fit in.

- This will then encourage investing both fixed and working capital, as well as time and effort, in the many and various measures needed so as to allow a good high immune platform in the breeding herd to contain the inevitable disease challenge.

If you must feedback . . .

If you feel you should – this is probably the most effective method of 'feedbacking'.

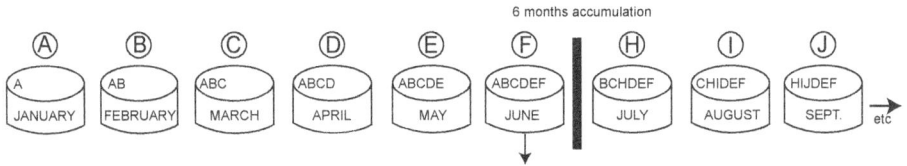

METHOD: A series of ice-cream cartons are held in a deep-freezer. Each month (or after a disease outbreak) SIX separate SMALL quantities of piglet scour/faeces, afterbirth, a dead piglet's gut and contents, are put into six pots and kept in the freezer. This is rolled forward for six months.

Once the sixth carton contains 6 months 'contributions', the pieces are slowly thawed, then minced and strained through fine mesh to provide a bacterial soup. This is slightly diluted with tepid water and watered on to the breeding pig's food. Continue rolling on with the current month's pot containing the past six months' contributions.

IMPORTANT: DO NOT DO THIS WITHOUT YOUR VET'S AGREEMENT. He knows about your disease 'picture' and it can inflame (not immunise against) some diseases (all feedbacking carries a calculated risk).

PMWS
(PORCINE MULTI-SYSTEMIC WASTING
SYNDROME)

Readers know only too well that PMWS is one of the most serious profit-sapping diseases to assault pig producers in Europe, and it is now spreading globally. What is thought to be the primary (there may be 'co-associated' agents) causal agent, a circovirus (Type PCV-2) has been in many of our herds for decades, and until a few years ago was not troublesome at all. Now something is causing it to flare up. What is that 'something'? Or has the virus mutated in some way? What are we doing, or now doing differently, which has changed things? My own country, Britain, was one of the first to be affected after France and Spain, and vets and farm advisers like myself were among the first, let's admit it, to tend to run around in circles not really knowing what would effectively combat the sudden disastrous post-weaning fall away in performance. Initially the veterinarians' reaction of high-level antibiotic treatment only mitigated the secondary infections, and attention turned to sprucing up management. Prof Madec issued his well-constructed '20 points' (*Table 1*) in 1999 which certainly helped when producers gritted their teeth and adopted at least three-quarters of the twenty points. I certainly got on to my own clients to do as many of them as they could, but eventually this puzzlingly didn't work after 18 months of effort and on other farms it did after six. Could it be that certain factors were more important on one farm compared to others on a neighbouring one? Probably.

This is the article I wrote after 15 months close-up experience of fighting the scourge in the field in the UK and France, Prof Madec's home territory. I'd not change it today but at the end I add some further comments which 5 years trial and error suggest might be worthwhile addenda.

Both PMWS, and its partner-in-crime, PDNS, are not finished with us, especially in pig industries outside Europe. I'm sure other implicated factors will be added to the growing list we already have.

'It's an ill wind....'

But this infuriating and morale-sapping disease is certainly forcing us into being much better farmers and thus – in the end, as is now happening in Britain, France and Spain – this ill-wind is blowing the whole job of pig production some good, if painfully, and on some farms needing strategic restructuring – expensively..

Incidentally, I am not surprised recent textbooks don't cover the disease all that well. This is no fault of their authors as the problem is complex and seems to be still developing. What a textbook could say today could well be revised tomorrow. So I would expect authors to be cautious. As for myself, in this book, I'm prepared to have a go at summarising the PMWS problem at the time of writing. Some six years sharp-end experience of what seems to work on many, not just one, farm or research station may be worth consideration.

Read on and decide for yourself.

Table 1. THE MADEC TWENTY

A detailed examination of the risk factors connected to PMWS has been conducted in France by Prof. Madec. The following chart sets out the 20 main measures recommended by the French researchers for farms with a severe PMWS problem.

Maternity	1.	Empty the slurry tank or dung channels, clean and disinfect them.
	2.	Wash the sows and apply antiparasitic treatment.
	3.	Cross-foster piglets only in the first 24 hours; reduce
fostering		to a bare minimum and restrict to the same parity range.
	4.	Use adequate vaccination programmes.
Post-weaning	5.	Have small compartments (pens) with solid dividing walls.
	6.	Empty slurry tank/channels, clean and disinfect.
	7.	Stock pens at rate of 3 pigs per square metre at entry.
	8.	Allow 7 cm of trough length per pig.
	9.	Ensure good ventilation.
	10.	Maintain ideal temperatures.
	11.	Do not mix groups: one group (week's weaning) per room.
Grow-finish	12.	Have small compartments (pens) with solid dividing walls.
	13.	Empty slurry tank/channels, clean and disinfect.
	14.	Follow a stocking density of $0.75m^2$ per pig.
	15.	Attend to ventilation and temperature.
	16.	No pen mixing.
	17.	Do not mix groups of different ages.
Other measures	18.	Respect the flow of animals and air.
	19.	Practise good hygiene with interventions such as castration and injections.
	20.	Remove confirmed PMWS cases to hospital pens.

(Adapted from: Madec and others, 1999, Journal Recherche Porcine en France, 31, 347-354)

What is PMWS?

Infection of young pigs, usually 8-16 weeks old (often peaking around 9-10 weeks) with a porcine circovirus Type 2 (PCV-2).

PMWS = Porcine Multi-systemic Wasting Syndrome
(Multi-Systemic = many areas within a body.)

Nursery pigs become pallid with stunted growth, there may be diarrhoea and of the 7 to 10% (sometimes higher) of pigs affected, mortality can be very high.

There is a carry over into subsequent growth depression.

Another acute viral disease PDNS, Porcine Dermatitis and Necropathy Syndrome (Necropathy = degradation/death of body tissue) which is readily identified by skin lesions on older pigs, can occur before PMWS, but more usually PMWS seems to open the door to it.

At the time of writing there is no vaccine or treatment, so control must major on keeping the PMWS out of the herd, and lessening or shortening its impact if it arrives.

PMWS can last for 18 months before immunity is established, but by following as many of the protocols described here it can be shortened to 6-9 months or less.

Because PMWS symptoms can be similar to other serious diseases like African Swine Fever, as well as several less serious diseases, it is essential to get prompt veterinary diagnosis if an outbreak is suspected.

Why is it so prevalent today?

Many of the 'new' viruses attacking our pig herds are virulent, often resistant to treatment and seem to be on the increase. This is probably due to 3 main reasons.

1. Farmers are still using 'old' methods of establishing natural immunity in their sow herds which are insufficient to provide the sows, especially replacement gilts, with sufficient time to acquire natural immunity to these 'new' viruses which can then be passed on as maternal protective antibodies to their litters.

2. These new viruses can be difficult to kill with traditional disinfectants (Lysol, formalin, sodium hydroxide) at traditional dilution levels. An approved *oxidative* disinfectant like Virkon at *specific dilution levels* is essential to get through the virus' own protective shield.

3.   The new viruses love organic matter, which gives them added protection against the virucidal effect of an oxidative disinfectant. Failure to remove *all* faecal and fatty caked-on detritus can protect these new viruses and allow them to survive on housing surfaces, equipment, boots and transporter vehicles. Thus a degreasing (*farm*-approved) detergent (like Biosolve) at the advised dilution and 'cover' recommendations is an essential forerunner to the subsequent correct use of the right disinfectant.

It is because many farms have not updated their management in all three areas that diseases like PMWS, PDNS, PRRS, Swine 'Flu etc are gaining a hold on their farms.

Indeed, all these new diseases can be categorised with just one set of letters – TPFDs or 'Typical Pig Farm Diseases!' 'Typical', because typically these farmers have not updated their disease control methods and/or products used.

Preventing PMWS from getting in to your farm

•   Buy replacement stock only from herds clinically free of PMWS. Ask your vet to check status with the vendor's vet.

•   Ensure the transporter driver carries a Certificate of Vehicle Disinfection signed by the vendor's biosecurity supervisor. He should have fresh overalls and clean boots. Even so, do not allow him on to your premises/ close to your pigs.

•   Unload into a 3-day quarantine pen *on the farm boundary*. Observe x 3 daily; then if all is clear, move the new intakes to your specific induction unit (see below).

•   Only use Virkon (or an approved paracetic peroxide disinfectant) as your disinfectant, including all wheel dips, washes and foot dips at correct dilution *and replenishment* levels.

•   Ensure there is no unauthorised access to the unit.

•   Control rodents and especially birds.

Making it more difficult for PMWS to get established.

It is noticeable that PMWS and PDNS are prevalent on continuous flow farms, so:

•   Change to All-In/All-Out (AIAO), especially in the farrowing and nursery houses.

- Even so the AIAO cleaning and disinfection routine at terminal disinfection is out-of-date on many farms …

- Thus you must get the surfaces spotlessly clean before the disinfectant is used. This includes removable equipment and under-surfaces. (All slats should have a raised area/trapdoor per pen. *(If the Brussels bureaucrats, so keen on legislation, want a new law, then making it legally mandatory to have all underslat surfaces accessible would be a useful one!)*

   So you must use a degreasing detergent (Biosolve is a good one) before you disinfect and allow it to soak in (to crevices) for 20-30 minutes before disinfecting.

- You then must use a product equalling the technical specifications of Virkon at 1:250 dilution rate (for PMWS).

- You must allow the surfaces to dry before reintroducing pigs. A good idea is to use a kerosene space heater especially in winter.

- If PMWS is present in a herd, upper/out-of-reach surfaces must be fogged with Virkon.

PMWS virus can survive in your waterlines, especially header tanks. The tanks and lines must be cleaned with Virkon.

Building a better natural immunity to PMWS

- Study the latest advice on induction of replacement stock. (Read page 94 onwards, for example, in this book.)

- Consult your veterinarian for day-to-day guidance on the "challenge" protocol to use, as all farms are different.

- Generally either buy weaner gilts (grower gilts) or …

- Allow at least 6 – 8 weeks (seek veterinary advice) from gilt delivery at 90-100 kg to first service.

- Split this into a challenge period of about 2 weeks (seek veterinary advice) where items like bedding from the weaner unit is spread among the gilts and/or other measures used, eg vaccination (seek veterinary advice).

- And then a rest/consolidation period of 5-6 weeks (veterinary advice) where no challenges are given.

- Don't overcrowd, keep the gilts warm and happy.

• Don't allow the gilts to grow to their genetic potential – slow them down to 650-700g/day, no more. A nutritionist will design a special gilt-developer diet for you.

Dealing with an outbreak

Should you get PMWS on the farm, the following measures will help you get through it.

Of course not all of them will be feasible. Do as many as you can while the outbreak is with you.

• Generally speaking, at the start weaners from only certain litters (maybe 10-15%) will be affected. The following measures will curtail pigs from one litter infecting others:

1. Only use one injection needle per litter.

2. Fostering and multi-suckling will spread the virus.

3. Batching and matching post-weaning likewise. Best if litter groups are kept separated for 3 to 4 weeks post (24 day) weaning, *if possible*.

• Parvo, PRRS histories and ongoing EP respiratory infection make it easier for PMWS to spread. Indeed the presence of one or more of them may be co-factors. Discuss with your vet about strategic vaccination for these or other diseases.

• PMWS virus loves overstocking; stuffy kennels or flat decks shut up on cold nights; temperature fluctuation; and chilly draughts. Check, check, check!

• All circoviruses love faeces (and mud). Don't allow areas to get too fouled up in the farrowing pen or nursery. While an outbreak is with you, wash dirtied floors or parts of floors more often.

• All these 'new' viruses have long 'recovery tails' before natural immunity eventually squeezes them out (PRRS is a classic example). During this slow climb back to full immunity, secondary bacterial infections can get a hold, from E coli scour to 'greasy pig'. These, if not jumped on quickly, can delay the recovery tail still further (a catch-22 situation) so watch out for them and attack them promptly. Your vet will need to 'type' (ie identify) them and advise on the best drug, product and procedure.

All is not lost

The pessimists say, "There's not much you can do to combat PMWS". The above notes suggest there is – a very great deal!

Five years on...

Here are some further thoughts.

My clients who have prioritised the following measures have won through the quickest.

- Critically reassessed and updated their cleaning and disinfection strategies.
- Reconsidered and then reorganised their pig flow....

  First: by adopting AIAO;

   Then: batch farrowing;

  Then: Adopting partial depop (PD) *as routine* (see pp 231-236).

  Agreed, this takes time. So start immediately.

- Became much more aware of stocking density, reducing the current common levels advised by 10%, and during PMWS/PDNS by 15% or more.
- Been less in a hurry to get gilts served so soon.
- Became more aware of preventing/controlling mycotoxins which *could* be affecting the part of the pig's immune mechanism which may be the trigger factor?

  (In two successive autumns now, PMWS has flared up again on clients' farms when new-crop cereals are used. Is this the effect of mycotoxins – or some other 'new crop factor' at work?

- Overseas – slatted channels/sub-surfaces (very prominent outside Europe) are also cleaned, or more diligently cleaned.
- Overseas – PRRS seems to open the door to PMWS, as if PRRS wasn't bad enough!
- Make sure you are using selenomethionine at the advised rate as well as being up-to-speed with your Vitamin E levels in sows and weaners. High performance weaners are probably low in the level of Vitamin E modern conditions may demand.
- Do a regular stress audit on your weaners. See page 135. The best person to do a stress audit on your weaners is another experienced breeder/nurseryman called in occasionally. And ask your vet to home in on this subject when next he tours your pigs, rather than concentrate on spotting disease, as they have to do.

The farm tour - another pair of eyes

No, not the organised group tour, which does have its place in the mind-broadening scheme of things, but something I consider even more important. When I was managing a pig farm I would get another experienced producer periodically to walk our unit and I to walk his in return, every six to nine months or so.

This is the age of stringent biosecurity. Such a suggestion today is now greeted with a frown and a sharp intake of breath! Does it spread disease? I don't think so. Providing the visitor has bathed, changed into freshly-laundered clothes, uses your overalls and boots and leaves the car outside the pig unit boundary, the risk is minimal. I'd go so far as to say non-existent. I've done it professionally on visiting what might be some 3000 or more pig farms over 40 years of advisory work in 21 countries, and have never had so much as a whisper of an accusation of spreading disease. But I'm meticulously careful, sometimes to the frustration of my 'minder' on overseas farm visits, who has a schedule to keep to and hasn't yet encountered my insistence to scrub up and down between visits in those other consultants he has had to cart around!

To forego this very useful opportunity to look, listen and learn in a friendly one-on-one manner for the sake of over-hyped and currently fashionable anti-visitor paranoia, is self-defeating - but not if the correct precautions are agreed and followed. Please do it.

Observation

A good stockperson - a good manager - has to be observant. Picking up changes as they develop, and of course acting on them. Good observation varies across a vast range of possibilities - from noticing the way gut disturbance in the young pig can be picked up 6 hours or so before it actually scours by the way it lies in the (raised) semi-sternum position, to the latest water intake monitoring data pre-heralding an onset of swine 'flu, pioneered by the Farmex company recently.
So it is with the pig farmer friend called in to look at your pig management. His observation **will be tuned to a different observational wavelength to that of you and your staff,** and will often pick up anomalies for discussion between the three of you.

CLEANING & DISINFECTION

I find it quite extraordinary that none (except one) of the textbooks on pig production I've studied deals anything like adequately with cleaning down and disinfection, and even that one is – shall we say – insufficiently comprehensive enough in my opinion. Considering that more of our modern textbooks are (excellently) written by veterinarians, who of all people must realise how important hygiene is on the pig farm, it makes my eyebrows rise even further whenever even they fail to cover this subject adequately in their textbooks!

Any serious textbook on pig production should devote a whole chapter to the cleaning down and disinfection side of farm biosecurity. And another one to the many other peripheral items to do with the huge subject of biosecurity, which goes far beyond the confines of just cleaning and disinfection.

Why ?

We are being assaulted by new and virulent strains of viruses – I don't think any pig producer needs to be told that. And I do not mean just the CSF and FMD viruses which hit the national headlines recently, but those causing PRRS, PDNS, PMWS, Swine 'flu as well. And hard on their heels are a whole clutch of what I call 'Won't-Go-Away' bacterial diseases (they do go away after treatment but often reappear) like Strep suis for example, E coli scour and maybe Ileitis. You know the problem!

Complacency?

Buttonhole most serious pig producers (as I have) and they say "But we clean and disinfect pretty well", "We have a set routine and stick to it", "Yes, we are now AIAO" (All-in/All-Out), "Don't all disinfectants zap everything?" and so on.

When I ask a few questions on what they actually do, great gaping holes appear between what we are told by the experts is necessary (not 'advisable' –

necessary) and what producers are still doing on their farms (*Table 1*). Thereby lies the key – what they are *still* doing today despite much printing and computer ink being expended on telling them what to do and how to do it.

Nevertheless I don't think this is complacency. Complacency implies that farmers know what is correct but it is not done for reasons of time, or labour, or money or a failure to monitor things diligently.

Sorry, but it is more likely to be ignorance of what is needed these days!

Table 1 SURVEY REVEALED 9 OMISSIONS OR ERRORS

I've surveyed 105 pig farmers across 18 months. Here are the results.
- Three quarters of them used no detergent in their pressure wash-down
- 68% of these did not use a *hot* pressure wash, even so
- Four-fifths of those who used a detergent did not use a farm-specific detergent
- Only 2% troubled to monitor the effectiveness of their cleansing and disinfection procedure
- Of the 9% who were sampled with swabs after disinfection, none at all had levels
of   viable bacteria remaining at or below the target level of 1,000 viable bacteria
per sq   centimetre (6,500 per sq inch).
- 50% of those swabbed had over 5 million/sq cm viable bacteria from at least one swab
- 80% did not sanitise the water
- 40% did not *regularly* combat vermin
- 90% "only fogged after a disease storm", not as routine. No one fogged their loft space

Where the ignorance lies

- Failure to realise how much disease costs you in performance. Probably 0.3:1 food conversion from 7 – 100 kg and 4 fewer pigs sold per year in the breeding herd. Yes, that much!

- Failure to recognise that subclinical disease – the continuous effect of rumbling, low-level, largely invisible disease *possibly costs you more*, over a period of say 2 years, than the outbreaks of clearly visible clinical disease we all worry about and take action on when it happens.

- Failure to realise that modern pathogens are tougher, more resilient and more virulent than ever before and so need uprated detergents and disinfectants to combat them.

- Failing to clean properly before disinfection. Ever painted the outside of a house? I'm sure you have. Experience from previous disappointments tells you that it is the *preparation of the surfaces* which lead to a long-lasting effect, not so much the care in application or number of coats of paint subsequently applied.   In the same way, pre-cleaning before disinfection has gone by default. We don't clean properly so we end up not disinfecting properly, with disease the disappointment.

We now have better virucidal (anti-virus) disinfectants, but they tend to be neutralised by organic matter and fat deposits.

These new viruses have stronger protective biofilms around them – they've changed so as to be better survivors. The 'old' disinfectants, like the phenols and quats, aren't so good at getting through this protection. Newer oxidative disinfectants (peracetic peroxides) are more effective at the job, as well as being more biofriendly – a spin-off bonus so that pigs can even breathe them in at advised dilutions (fogging). They can also be used, with care, in the pig's drinking water.

However, they do tend to be weakened by organic deposits on surfaces. In addition, fat and grease on the piggery surfaces can make their job harder. The nutritionists are using more fats in lactating sow, baby pig and nursery diets these days.

You must remove these barriers to get a good kill of pathogens from the disinfective process – down to about 100,000 TVC (Total Viable bacteria Count) per cm^2 *before* disinfection. To think that you can start with 50 million/cm^2 and only reduce this to 20 million/cm^2 if you just use a cold pressure wash before applying your disinfectant wash (Waddilove 1998), you can see the problem you are giving any disinfectant!

Sorry guys, but to get down to the necessary 100,000/cm^2 before disinfection so that the disinfectant can lower the level of TVC to well under 1000/cm^2 – a hundredfold decrease – a level which the pigs immune system can handle (no farm building is ever made sterile, nor need it be) then you've got to use a degreasing, farm-specific detergent, preferably applied with a *hot* pressure washer and allowed to soak in for 20 minutes before you use your disinfectant.

"But the COST, Mr Gadd!"

"Wow!" You say, "That's going to cost a bomb in products, time and extra gear!" Sorry, it is, but the payback (REO) from better performance is rarely less than 5:1 and can be as much as 12:1 (*Table 2*).

Recent surveys/calculations by Antec International, part of the DuPont Group, suggest it is 7:1. I'll settle for that!

Standard industrial detergents used to clean factory floors are not good enough for our farm buildings these days. Use an Approved, farm specific detergent at the advised dilution and coverage.

For disinfection, there are a range of modern products. They all cost more per square metre covered, some of the best virucides more still, so you must seek advice from veterinarian and manufacturer as to *what* you need, *when* you need to apply it, and *how* to use it, i.e. using the appropriate *dilution rate* and *coverage allowance*.

So… better, more expensive products but used correctly – all this is bound to cost more. Is it worth it? Tables 2 and 3 give pointers as to the cost-benefit picture, based on research and farm trial results, and modern costs of the products required and the equipment needed to apply them effectively.

Table 2. THE BENEFITS FROM BETTER BIOSECURITY TECHNIQUES

Reference	Trial type, basic details	Calculated value of finishers (against controls or former practice)
Cargill & Benhazi (1998)	Cleaning AIAO buildings before disinfection with a detergent	+ £3.12 /pig ($5.68, €4.59)
Overton (1995)	Salmonella outbreak controlled	+ £3.86/pig ($7.02, €5.67)
Jajubowski *et al.* (1998)	Using a peracetic acid disinfectant instead of NaOH	+ £8.80/pig ($16.02, €12.94)
Sala *et al.* (1998)	Full Antec programme v. iodine	+ £2.10/pig ($3.82, €3.09)
Sala *et al.* (1998)	Full Antec programme batch disinfection v. terminal disinfection only	+ £5.66/pig ($10.30, €8.32)
NCASHP Denmark (*nd*)	Partial v. total biosecurity programme	+ £7.77/pig ($14.14, €11.42)
Antec Trial (G&M, 1999)	Change to AIAO and updated disinfectant, result after 3rd batch	+ £7.15/pig ($13.00, €10.51)
Gadd (1994-1998)	Average of 10 clients uprated to full biosecurity protocols	+ £5.63/pig ($10.25, €8.28)
	Average : all results	**£ 5.51/pig ($10.02, €9.00)**
Assumptions:	Weights ranged from 6–90 to 30-100 kg. Food in last 14-21 days, range 2.2 to 2.25 kg/day. Finisher feed price £130 ($237, €191)/t KO% standardised at 73%	

What you should consider doing

1.　Review your cleaning and disinfecting protocol. Go back to school on the subject and get updated. Get the literature from a reputable *farm* hygiene firm with a good track record or ask their rep to call.

2.	Get your veterinarian on board. You'll need him to help you identify the viruses lying in wait on your farm or in your immediate locality, and advise which of the grades of bactericide/virucide you need to use. Some need more sophisticated products than others.

3.	Follow his and the biosecurity firm's advice. Please don't argue! Sure, it will involve you in more work, more cost, more equipment – but the payback of 5:1 and which has been up to 12:1, even with all those extras, is just too good a bargain to deny yourself.

4.	Check, check, check that your staff are doing the tasks properly and that the standard isn't falling off. This is a natural reaction to any dull and routine task. There are tests kits available. Use them as a normal part of management, just like a thermometer to monitor environmental temperature.

Table 3. PROPER BIOSECURITY – THE COST PICTURE (Figures based on the surface areas/ labour and materials needed to prodce 100 finished pigs on a farrow-to-finish farm, including all breeding unit surgace areas needed for 4.33 sows) Conversion rate: £1 = US$1.82, €1.47

Expenditure on materials	Correct biosecurity protocol	What you do now	Extra cost
Cleaning materials	1.07p – 1.3p/m²	No detergent used	1.07p – 1.3p/m²
Disinfection	0.79p-1.7p/m²	0.61p/m²	0.18-1.09p/m²
Airspace fogging	38p per 100 pigs	Rarely done	0.38p/pig
Water sanitation	£5.43 per 100 pigs	Rarely done	5.43p/pig

Labour (& Cleaning Equipment)

Cleaning	No difference		–
Hot pressure wash	Cost of steam jenny over 8 years £1.60/100 pigs		1.6p/pig
Disinfection	No difference		–
Airspace fogging	£1.30 per 100 pigs		0.13p/pig
Water sanitation	65p per 100 pigs		0.65p/pig

Total extra cost per pig is as follows:
At 15m² total surface area needed to be sanitised per finished pig (including the breeding farm) and taking the most powerful/costly products available at the more concentrated dilution rates at 2.39p (4.35cUS, 3.5c€)/m² the extra cost per pig is 15 x 2.39 = 35.85p ($0.65, €0.52). Add to this the remaining extra costs/pig listed above of 8.19p, the total extra cost over what is done is now 44p ($0.80, €0.65)/pig.

Note: UK costings and metric measurements are used here because (see below) we are only comparing costs against likely return.

The benefits of doing it properly

Table 2 is based on 17 comparative trials (11 farm and 6 research) where on 16 of them the pigs were not suffering from any particular or obvious disease.

This is important, as good biosecurity should lessen the incidence of clinical disease outbreaks, where the benefit would be much greater – possibly up to 3 times as much?

The cost of doing it properly

The additional cost of proper biosecurity is made up from *the extra cost* of the modern materials over what you use now (i.e. *virucides* rather than just bactericides); the costs of *extra tasks* now considered important (ie *hot* pressure washing using a heavy duty detergent; *fogging* the airspace and *sanitising* the drinking water) and the cost of the *extra labour* all this entails. Table 3 is an attempt to draw it all together.

In terms of Euros, the 44 pence sterling converts to €0.66/pig. If from the trials cited in Table 1 the expected benefit from a complete biosecurity protocol is €8.27/pig then the REO (Return on Extra Outlay) is €8.27 ÷ €0.66 or 12.5:1. When a good growth enhancer typically obtains 6:1 REO at best, this puts proper biosecurity into its true perspective – a very good bargain indeed.

Recent figures from a major manufacturer reveal, by my calculations on their recent product costs, a REO of 7:1. This is still a rattling good return – run it across the nose of your bank manager!

Doing it properly – the HACCP Route
HACCP (pronounced 'Hassap') = **H**azard **A**nalysis and **C**ritical **C**ontrol **P**oints

Table 4 HACCP BIOSECURITY – 7 PRINCIPLES AND BRIEF EXPLANATORY NOTES

Hazard Analysis	Identify the hazards (e.g. bacteria, virus or fungus) at each step of the process (e.g. transportation or personal hygiene)
Critical Control Points	Point in the process where the hazard can be reduced or eliminated (e.g. foot dip)
Critical Limits	A defined reduction for the hazard at that control point (e.g. Campylobacter – 99.999% reduction in the number of organisms in the environment)
Monitoring	Observation and measurement to ensure procedures meet critical limits (e.g. swab tests according to statistically significant criteria)
Correction	Action required if measurements fall outside critical limits (e.g. review application procedure)
Recording	Records must be kept of all limits, products and actions for control and legal reasons (e.g. daily cleaning schedule)
Verification	Outside measures to make sure the whole system is working acceptably (e.g. third party to check dose rates or that all records are complete)

VEHICLE SANITATION

I was caught up in the unhappy saga of our Foot and Mouth disease outbreak across 2001. Any of us severely shaken Brits involved learned a whole pile of do's and don'ts for the future, and a crop of official 'enquiries' have taken place in order to drive home some of the critical lessons we have learned – or I hope we have learned.

I'm going to pre-empt most of them and come out with some of my own opinions about what any pig industry has got to do to keep any disease down, let alone one as contagious as FMD.

These guidelines will not generally be found in any current textbook, but I guess they need to be in future!

Vehicle movement as a primary disease vector

Responsible people among us in Britain are now agreed that vehicles are a major cause of livestock disease spread. During the FMD outbreak – even well into the outbreak – I saw with my own eyes a whole crop of errors. Inadequate wheel arch and underbody disinfection. Inadequate replenishment of farm-gate disinfectant. Disinfectant pads less than the circumference of the vehicles' wheels – and so on! I banged the drum in the press and got unpopular with livestock farmers already under great strain but, by Jingo, it needed saying!

I have therefore drawn up a set of suggested procedures which both livestock farmers and those who supply and collect pigs, food and supplies from their production units should follow if we are to really reduce the very likely threat of one farm infecting another through a third party. This is what I am telling my clients they have got to move towards from now on.

Controversial? I don't think so even if they will cause much argument. Difficult to carry out? In a way, yes, as it will involve persuading that (or insisting) other people should co-operate. Costly? More costly, yes, but look what disease is costing us at present. And I don't mean FMD (which cost *every man, woman and child in Britain*, including consequential losses, about €200 (£136, $248) but diseases like PRRS and PMWS which can run away with most of your profit for 6 months or more and a third of it for at least another year.

Suggestions

- Farms should have *perimeter* unloading and loading facilities for pigs, feed and general goods. No delivery/collection vehicle, including salesmen's' cars or even the vet's vehicle, should be permitted past the farm boundary, or alternatively cars only allowed into a biosecure area with separate access on to hard standing with cleaning facilities attached.

- No pig collection or delivery vehicle should be accepted without a form/ certificate logbook issued by the buying organisation/vendor, signed by the buyer/vendor's biosecurity supervisor that the vehicle has been properly cleaned and disinfected – inside, under and outside– before calling at the farm. These organisations to train and appoint a biosecurity clearance officer as routine, with management supervising his role, too.

- In addition, each vehicle must have available a written load record which the producer can examine. This is to deter the knock-on effect when visiting a succession of farms on one delivery/collection journey. If you have to accept a multi-delivery vehicle *do not allow it on to your farm*. Provide a covered off-load point on the perimeter and insist that the delivery is telephoned in by the driver and go and sign the docket there. Everyone has mobile phones these days, so there is no excuse for it being 'unworkable', as I've been told. Tell your supplier to get into the 21st century; pig diseases are, even if he isn't!

- No driver is allowed to assist in loading/unloading pigs.

- Anyone driving a transporter must not be allowed *by law* to keep pigs on any other premises.

- All abattoirs, markets, feed mills and commercial pig-breeding companies must invest in satisfactory tunnel-type disinfecting bays with a supervisor attached to ensure thoroughness and avoid 'dodging' when behind schedule.

- Farm pig loading bays are essential, with wash-down facilities on tap. Drainage should be away from the farm and the access road to and from the loading ramp should be a dedicated route, not used by everybody.

- Casualties must be incinerated on site. In certain circumstances they can be left for collection e.g. rendering etc at an off-site venue, suitably protected from exposure and degradation, with the site disinfected after removal.

- Wheel dips on routes to and from the farm are a wise precaution but must be adequate i.e. with automatic or manual spraying of wheel arches and undersides of all vehicles passing through. Such devices are now on the market. Insist on them, as disinfectant pads are inadequate.

- Use strength dilution monitoring tests for disinfectants (available from the better manufacturers) to keep abreast of contamination by organic matter. Even the most effective disinfectants are weakened by mud and dirt.

- Place clear signs at perimeter exit/entrances, and provide your suppliers of goods and services with specific instructions on biosecurity-related measures you will expect them to follow.

- For bulk feed deliveries it is preferable to have your own off-loading blower hose, as many truck-mounted hoses are dragged across farmyards and rarely disinfected for fear of contaminating feed and for delaying deliveries. Any filters should be duplicated, dismantled and cleaned between farms.

How vehicles spread disease

Since FMD, and now on salmonella spread, some interesting work has been done on delivery vehicles as disease vectors. This found that…

- Wheels are less important than wheel arches, mud flaps and the whole chassis underside. These (as we know) collect organic matter in which pathogens breed.

- Double-wheeled vehicles are a problem to sanitise. In future built-in disinfectant sprays may be needed, above each out-of-reach wheel space.

- Cabs are a disease vector, lessened by the driver wearing protective overshoes (to be left behind on the farm), and having machine-washable footpads in the cab, and to carry a portable vacuum cleaner.

- Tarpaulin covers are a real hazard.

- Washing two or more vehicles together on the same bay risks spreading disease between them.

Rather than regarding visiting vehicles as a welcome arrival – *consider them as a threat and take defensive precautions*.

And finally ...

Something really controversial in my own country.

Phase out all livestock markets! Before they were closed, these spread FMD like wildfire in 2001 and once re-opened as many have been (pandering to ingrained local custom and not science or even commonsense) constitute a disaster waiting to happen again. They *must* spread disease, even at the best of times.

In Britain and in some other European countries few pigs are marketed this way in contrast to our medievally-minded cattle and sheep farmers! If you are one of those still involved with the archaic live-auction market system I'm sorry to upset you – but you may as well play 'Russian Roulette' with your pig venture in future. Don't do it: let the other chap 'come a cropper', as we say in England.

Trouble is, the diseases he catches (and spreads) from buying weaners or in-pig sows at market, could find their way into your unit too.

HOW TO BUY SANITISING MATERIAL
(DETERGENTS & DISINFECTANTS)

Pig farmers are busy people, and they are exhorted to learn about so many things these days (nutrition, breeding, disease prevention, rules and regulations, ventilation, insulation and thermodynamics, etc) that the brain tends to take a rest when it comes to having to re-learn about cleaning and disinfection, a boring old routine subject which has been with us for ages.

If the product has a good name to it, is recommended (particularly if it is a familiar old favourite used for years), the tendency is to buy it again – and again, and get on with it! But things have moved on. In this field too, like many others, there are a wide range of products which are most effective when used to protect against the disease picture currently on the premises – which in itself can change, as you well know.

Additionally, you would not (need to) use the same disinfectant with which to wash your hands when the disease situation can be considered 'normal' on the farm as would be essential for general disinfection when a serious virus disease is present in the pig pens.

Here are some ideas on what to look out for when buying detergents and disinfectants cost-effectively.

How to choose a good detergent

It must be farm approved. What does this entail?

1.   Capable of working well on all surfaces found on a pig farm. Unlike urban factories, there are many *very* different kinds of farm surfaces. Several of them are semi-porous (e.g. concrete, plastic and some metals). This variability makes it more important to use a product specifically designed for on-farm use. A heavy-duty formula is essential, ***stronger than those used in a catering establishment***, for example. In my survey 18% of the farms used a well-known catering detergent 'because it was cheaper'.

2.   Contamination in crevices and other poorly accessible places is more easily removed with a heavy-duty formula.

3.   Slats are more thoroughly cleaned. The build-up of dung on the surface *facing between the slats* is more easily dislodged. This is especially important with enteric organisms such as *E. Coli* and *Serpulina hyodysenteriae* (Swine Dysentery) and *Lawsonia intracellularis* (Ileitis).

4.   Good degreasing is vital. Just because a surface looks clean it does not mean it is clean of all pathogens. The presence of a greasy layer on the surface increases protection of micro-organisms by long chain fatty-acid molecules. A heavy-duty alkaline formula helps remove this protection; the alkalis breaking it up quickly. This is important, as the newer, essential and better virucides don't work so well with fat protecting the organisms.

5.   Vital if time is limited – a heavy-duty detergent works quicker and faster.

6.   It mustn't interfere with the subsequent disinfectant's activity. This highlights the importance of using a fully integrated programme, such as the Antec Pig Biosecurity Programme, when the products are specially chosen to be compatible, or in some cases help each other.

7.   Ideally it should be applied through existing equipment with minimal modifications.

8.   Foaming can be helpful. This increases the contact time and allows operatives to see where it has been applied. The foaming decreases the amount of water needed in the soaking and pressure-washing phases of cleaning. Reducing water reduces costs and problems with excess run-off to dispose.

9.   It does not leave residues that can make the floor slippery and harbour micro-organisms. Especially, it should not leave cumulative residues.

10.   It should work in hard water situations.

11.   It should be non-toxic to pigs and operatives.

How to choose the right disinfectant

We should now have a clean and exposed surface with a TVC (Total Viable Count) of 100,000 cm^2 or less bacteria. Do you (or your vet) check on this periodically? You should do. Such a bacterial threshold should also bring down viruses to a controllable level. The *surfaces* are now ready for disinfection, but we also have water tanks, lines and drinkers harbouring pathogens, and pockets of air, such as in lofts, which need attention to prevent recontamination. The problem with the older disinfectants has been that …

- They are poorly effective against some of the newer viruses unless used at impracticable and costly concentrations.

- Due to toxicity they cannot be used in water lines or as space foggers.

- There is a wide range of correct dilution rates and coverage areas to deal with certain pathogens. Stockpeople risk getting them wrong and owners order up the wrong disinfectant basing their decision on price.

- Some are toxic or irritant.

A disinfection check list

1. There is a wide range of disinfectants on the market.

2. You must choose one which is Approved for the disease spectrum you and your veterinarian are likely to encounter.

3. So either take advice from your veterinarian , or only buy from a well-known primary manufacturer who can advise you on which one to choose.

4. Equally important as the choice of disinfectant is to follow the Approved dilution rate which will differ according to the disease situation. Some manufacturers have a simple colour dipstick (e.g. Antec) so that you can *quickly and easily* check that the product is correctly diluted for the purpose in mind. BASF have one for drinking water sanitation.

5. Also important are the instructions as to cover rate. Most people either use a 'chisel' pressure washer at 200 psi or a spray nozzle. Managers should check the usage rate of the disinfectant purchased against the surface area which should have been covered in, say a month's use.

6. Check the time rate recommended for the disinfectant to act fully.

7. Some disinfectants take longer to work in cold weather. You may need a higher concentration, so check with the manufacturer in cold weather conditions.

8. If in doubt or you are unwilling to go into all these careful details, just consider using a peracetic acid disinfectant, the best known are Virkon or Virkon S. These are powerful oxidising agents and can rapidly kill most viruses as well as all bacteria; especially if you clean down well and then follow the instructions to the letter.

New areas

The water supply and the 'difficult-to-get-at' surfaces (ceiling space; under the slatted surfaces and underslat channels) are important areas where bacteria and viruses can initiate or prolong a disease outbreak.   PMWS has taught us that.

There are products which can be safely used in the water system and you need to know what these are and how to use them.

Respiratory organisms can often be effectively, and again safely, controlled through 'fogging' agents; this is another fertile field to explore. This can even be done with pigs present – seek advice first.

In summary:

•   Buying biosecurity products is now a skilled process.

•   Consult your veterinarian regularly. The disease picture changes.

•   Dialogue with the manufacturer. There are 'horses for courses'. Products, usage rates, cover rates and timescales differ. Check, check, check.

•   Before worrying about cost per litre or per can, always relate it to the dilution and recommended cover rates. You may get a pleasant surprise with some of the 'expensive' products.

Modern techniques
and
new possibilities

ESSENTIAL RULES FOR SUCCESSFUL 28/30 DAY WEANING

Further restraints within the EU are being put in place on the use of AGPs (Antibiotic Growth Promoters) and on cheap and convenient gut conditioners like high level copper sulphate and zinc oxide. To combat these assaults I show on page 219 that weaning at 28/30 days in place of 21/23 days, far from being a financial burden, promises improved margin.

However there is a strong proviso. This is that the sow, faced with a final week of suckling from 8.5 kg giants (rather than 6.5 kg piglets as of now) risks being dragged down in condition to such an extent that her subsequent reproductive performance is compromised.

Thus the whole economic success of later weaning – which field trials have verified – depends on:

- Skilled and effective creep feeding of the sucklers

- Defending the sow's condition across lactation.

The writer has been involved in farm trials on various methods of creep feeding for over 30 years, and has 20 years experience of the problems which breeders in hot/humid climates have with sow condition where the problem of condition loss is commonplace. Having just written a textbook where some 50 pages are devoted to these two areas alone it is difficult to deal adequately with either in one short article, so here are the key points as I see them. The things *you must* do, before all else. In this essay I deal with creep feeding – the next one the sow herself.

Creep feeding

- **Formulation not specification.** Choose a good creep feed from the raw material point of view. It is the *ingredients* which matter, not so much the specifications, as most pig nutritionists know what are the nutrients the piglet needs to parallel sow's milk. But if the raw material matrix from which they come is wrong – even if the digestibility is right, the piglets won't eat it, or enough of it, so where are your excellent specifications then! No, I don't mean palatability, although that is important too, but special ingredients which *favour* the neonate's gut lining, and not aggravate it, like certain types of soyabean do, for example.

- **Freshness is vital.** Order small quantities, never longer in your store than 14 days, and get an idea of when it was made from the manufacturer not the warehouseman; 3 days is maximum. Yes, a 'use-by-date' is desirable, but I doubt if many manufacturers will accede to this voluntary discipline. And creep feed must be stored cool such as in an old insulated ice-cream delivery vehicle. And never where pungent odours can be absorbed.

- **Texture is vital.** Correct pellet size and hardness – get expert at this. There should be no 'fines' at all – our trials when I was with the feed firm RHM showed even 10% fines reduced creep feed uptake by 50%, even with a 'good' creep feed in all other respects.

- **Cleanliness is vital.** Baby pigs are messy eaters, just like human babies! Like humans, the feed receptacles need to be kept spotless, which involves a lot of attention and elbow grease. It is one of the most important tasks a stockperson can do, so *make time* to *do it*.

- **Clean water fairly near to the creep** is essential to start with, then move it to the wet-end of the farrowing pen. The Japanese have these lovely little stainless steel leaf drinkers, and get good creep uptake. But they are also naturally clean stockpeople and take a lot of trouble. Sure, good creep feeds increase thirst, but this is not a problem if watering **is *good watering.***

- **Creep feed location.** Well-lit (but not 'well-heated'), nearer to the sow's head and out of her reach when lying down.

- **Others.** Liquid or semi-liquid, ie 'porridge'? Works well if the receptacles are kept clean and sweet (difficult).

  Even with a good creep feed, uptake will always vary markedly between individuals. Baby pigs will detect moulds, disinfectant splashings and over-mineralisation of the creep feed from 100 paces!

- **When to start?** Gently within 7 days – but little and often, even if it is not consumed. Remove uneaten creep feed (after 6 hours) and give it to the sow; she'll love it and anticipate the treat which lowers her stress/boredom factor. Patience and repetitive work is needed to get baby piglets started well.

- **Nutritional looseness?** The creep is probably not digestible enough, or contains too many gut irritant raw materials. Refer it back to the nutritionist/ formulator who will probably say it is bacterial! It won't be if you keep things spotless. Pay the extra for a reformulation improvement.

- **Finally, expense.** The difference between a superbly formulated creep feed and a 'reasonable' one in price terms can be a factor of 2.6 times more per bag. Ouch! This does not matter, as the piglet eats so little compared to its total feed consumption to slaughter, which even at a frightening creep feed cost per bag is still only pennies more in the total feed cost per finished pig.

I deal with defending the sow against heavyweight sucklers on page 213.

But is later weaning for you?
Yes, if …

- You have a post weaning check of more than 4 days. The 'postweaning check' is the time the newly weaned takes to regain the growth rate per day which it enjoyed on the day before weaning, hopefully 200g to 250g/ day at 21 days, for example.

- You are bothered with postweaning looseness. Sure, check up on your cleanliness and quality of your postweaning feed, but that extra 1.25 – 1.50 kg body weight and stronger physiology when weaning later/heavier can tip the piglet defences across into coping with it.

- Your stock people are not spending enough time caring for the pigs across the weaning period. Of course, get that rectified too, but later weaning can help with less skilled stockmanship while you do.

- This is as yet not proved, but if you are beset with virus diseases like PMWS, there is some evidence that later weaning can help. No, not necessarily, if …

- Stockmanship, quality of food and environment are of the highest order.

- You are unskilled at creep feeding.

- You are not good at lactation feeding of the sow (getting enough food into her).

- You haven't (yet) got the pig places available in the farrowing house.

So get these last three sorted out anyway and rea-assess the situation.

And what about the economics . . .?

Table 1. THE ECONOMETRICS OF 23 DAY V 30 DAY WEANING

	23 days	*30 days*	
Gross margin / sow (indexed)	100	116	16% better
Return on breeding farm capital (indexed)	100	110	10% better
Feed costs (€) ⎫	40.35	33.92	10.6% lower (ie better)
Fixed costs (€) ⎬ per finished pig	76.25	69.68	8.7% lower (ie better)
Total costs (€) ⎭	114.18	103.50	9.35% lower (ie better)

Source: SCA (2002)

More economics

I have been redesigning some UK farms to accommodate '8.5 kg weaning' rather than '6.5 kg weaning' as I prefer to call it – what are the increases i.e. increased housing costs likely to be in relation to the feed and fixed cost reductions?

Table 2. THE ECONOMETRICS OF LATER WEANING FROM BRITISH EXPERIENCE

Housing costs		*Performance costs*	
Extra farrowing requirements	+ 17%	Reduced sow output/year	1.04 pigs
Fewer gestation places	– 3.3%	Reduced gross margin per finished pig	– 4.0%
Fewer nursery places	– 4.4 %		
Net cost	+ 9.3%	At a gross margin of €15	€0.74/pig
Total penalty from both of the above		€1,07/pig	
Total benefits from Table 1		€10.68/pig	
REO therefore 10.68 ÷ 1.07 = 9.98:1			

These confirm and even enhance the indexed economic value quoted by the SCA work in Table 1.

Of course Table 2 figures are based on EU costs and you will need to put in your own figures – but it looks like being a worthwhile exercise in any country as the payback is so high.

LATER WEANING – HOW TO DEFEND THAT SOW

Previously I provided some comforting evidence that the proposed move within the EU to delay weaning to 28 days – far from being a burden to profit, could in fact be a bonus. Especially once an EU ban on gut-conditioning post-weaning feed additives arrives (all AGPs, long-term high level copper and short-term zinc oxide addition).

Another problem that 28-day weaning will bring is the considerable drain on the lactating sow from having to nourish a litter of 8 to 9 kg sucklers in place of those weighing 6 to 7 kg as of now.

On page 209, I provided a 'must-do' list of factors surrounding effective creep feeding of these heavier litter members.

Despite this, the sow has got to be protected from the enthusiastic assault of her very much bigger babies in that critical last week, when, say, 10 piglets could be growing at an average of 200g/day. On a basis of 4g of sow's milk being needed for every 1g of piglet gain, such a litter would demand 8 kg (10 piglets x 200g x 4) of milk a day, or 8 litres, near enough. 56 litres in that last week! (Next time you pass by a supermarket's fresh milk counter, have a look at the volume taken up by 56 one-litre cartons – pretty sobering!)

Experience

My experience is that even if one is good at getting creep feed into piglets early enough and sufficiently enough, such a potential demand still risks the sow being thrown into a last minute 'nose-dive' in body condition. I suspect that there are two types of nose-dive condition-loss in sows as Figure 1 suggests. The former steadier & progressive loss in flesh and fat cover down through lactation has been recognised as the most common up to now, but the second sudden last-third of lactation drop in condition may, in my opinion, have just as serious an effect on the sow's regenerative capacity to switch effectively from lactation/litter-feeding mode into full rebreeding mode, which involves a very different cocktail of circulating hormones across only a few days.

This delay affects days to conception (say from 5, to 7 or 8) and while this doesn't look much on paper, it could negatively affect subsequent litter size, as Canadian researchers have implied.

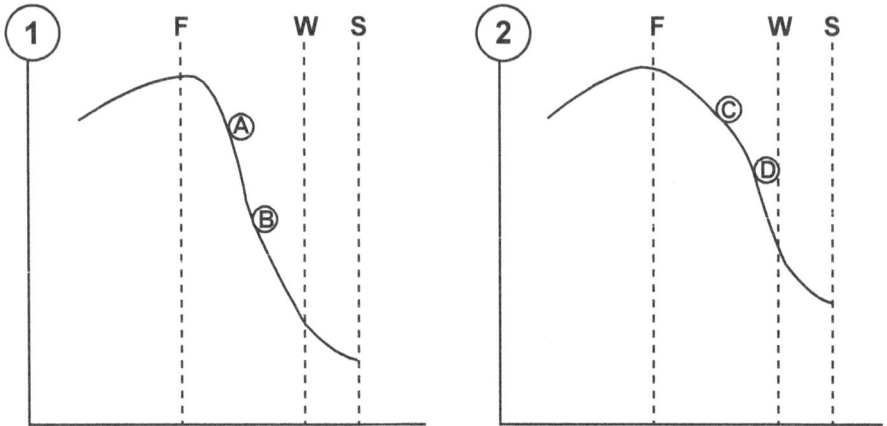

Are there two types of condition-loss in lactation?

1. Classical: Condition-loss is considerable, hormones unable to switch back sufficiently before service at 'S'.

2. Postulated: Condition-loss overall is less but the ***speed of loss*** is much quicker toward weaning, which could result in exactly the same effect in sows which don't look or feel so thin.

Key: F = Farrowing   W = Weaning   S = Service.

(A) & (C):   Time condition-loss is felt with the fingers.
(B) & (D):   Time condition-loss is seen by eye. By now it is too late to do enough to counteract the loss.

Defending that sow – a checklist

These are my priorities, based on experiences under tropical conditions in the Far East. Why hot weather? Hot weather affects sow appetite the most, so successes I've had in boosting sow nutrient intake in lactation despite such appetite-depressing hot and humid circumstances out there in the tropics could also be guidelines on what we might do to help any sow, anywhere, to boost her food intake when larger weaners are to be the norm in future.

1.   Long-term, select females with as large an appetite potential as you can.
    Different strains (even within the same breeding companies) *do* seem to

eat more than others. We will need to seek them out. And the breeders, may I dare to suggest, might have to choose rather more 'female' females than the 'male/female' proportions they do now – done with an eye on the important mainly male-influenced slaughter pig traits? However, later weaning could change this emphasis in the future breeding gilt and sow.

2. *Feed wet, preferably by pipeline.* Certainly in hot weather this will increase daily intake (range, in my experience, from 0.5 go 1.8 kg/day). This is a material contribution in deterring the sow from milking off her back.

3. ***Keep the lactating sow cooler.*** Easier said than done, especially in summer when the threshold is 20-21°C dependent on the sow's weight and condition, but the number of farrowing houses I enter where the sow is far too hot (4° to 6°C above this guideline) and where we have managed to get the sow feeling much happier ***and eating more*** by adjusting air positioning (plus a few other dodges – drip cooling and snout freshening) shows a success rate of 4 out of 5. Farmers are going to have to devote more attention to this area with 28/30 day weaning and modern fast growing sucklers.

   To find out the nuts and bolts of how to do this you'll need to read my own textbook as there are many management suggestions on this subject to be considered.

4. ***Sow comfort.*** A sow which feels good (*'The Feel Good Factor'*) will eat more, lose less condition when having to feed the bigger sucklers of the future. Fully half the floors in the farrowing houses I visit leave much to be desired – leg discomfort and foot lesions are too prominent. We need more research on crate flooring – visit any big Expo these days and you will see up to 30 varieties on show – and out in the field another 100 more on farms, with some floors which I'm afraid are quite awful when wear over time is added to poor design.

5. ***Taking the weight off the sow.*** Fostering, etc, when done well is fine, but the jury is out on this idea as, done too late, it could inflame PMWS, etc. On the other hand split-weaning, or as I've always advocated 'Weaning by Weight, not Date'[1] fits in well to Weight Related Segregation if not Age Related Segregation strategies.

6. ***Nutrition.*** Any nutritionist can design you adequate nutrient lactation feed specifications – just be agreeable to pay for them! Even so, palatability

---

[1] Something the EU legislators may not have considered?

has always been important, and deliberate feed fermentation is a new area worth exploration – part of the wet feed concept But the old advice of gradually increasing food offered over the first week (do follow the 'Stotfold Scale', page 214) and not overfeeding in pregnancy is just as vital as it ever was.

7.   ***Lastly*** – water. But by no means least! Fresh, cool, *'sucked up quickly water'* in my opinion, is better than a drinker, so a good separate 'elephant water trough' will assist food intake. If you do have to use drinkers, check the 2 litre/min flow rate *and make sure they are conveniently accessible* -1 see so many that must give the sow a crick in the neck (if she gets such a thing!), which cannot be good. I discuss this on page 155.

Defending the sow against a large or later-weaned litter – a checklist

Four critical areas.

1.   ***Preparing her for big litters.***

2.   ***Boost lactation feed intake.***

3.   ***Keep her cool.***

4.   ***Take the pressure off her.***

Preparing her . . .

•   Don't breed gilts too soon (<220 days), too light (<125 kg), too thin (<18mm P2).

•   Buy genes with good appetites.

•   Pay particular attention to the first litter sow – don't let her lose condition.

Boosting lactation feed intake . . .

•   Feed a special lactation diet (altered for hot conditions – a specialist pig nutritionist can advise on this).

•   Adequate water *and accessibility*. (Bowl or trough, not nipple drinkers).

•   Feed wet by pipeline; slightly fermented feed can have a dramatic effect on intake.

- Don't *over*feed in pregnancy, especially 7 days before farrowing.

- Feed must be fresh, smell wholesome, no mycotoxins, and kept cool.

- Feed 3 times a day, last main feed at night.

- Gentle downdrop of outside air over feed trough delivered by a special tube or pipe.

- Remove stale food (>12 hours).

- Follow the 'Stotfold' lactation feed scale guideline (overleaf).

- Give her (clean) uneaten creep feed.

Keeping her cool . . .

- Keep below 21°C/70°F in lactation.

- Avoid stuffy stagnant air ('Airbag' air placement in hot conditions).

- Place creep heaters in a covered creep.

- Consider neck water drips in excessive heat.

Taking the pressure off her . . .

- Foster/piglet swap (but take care if PMWS is about).

- Wean by weight, NOT by date. Practice being flexible.

- Stress is any pressure on a natural behavioural pathway. Discomfort is a major lactation stressor.

- *Keep boars out of sight, out of smell, out of mind.*

Body condition scoring has its detractors among academics, due to its undoubted imprecision and subjectivity. However, many years of advisory work has convinced me that its value lies in encouraging stockpeople to really examine their sows by feeling as well as looking. It is especially valuable in monitoring the start of the nosedive phenomenon in lactation. Academics, while right to point out its disadvantages, should ease up on their public denunciations for fear of damaging a useful sharp-end tool.

The 'nose dive' is a major problem worldwide, and anything which directs the pig technician's attention to preventing it - or even ameliorating it - can only be beneficial.

THE STOTFOLD LACTATION FEED SCALE (matrix)

First 10 days (All sows/gilts)			Sow Identification											
Day	Kg	Fed	Total Fed:											
1	2.5		Date Farrowed (Day 1)											
2	3.0		NOTES: *This dietary scale is now widely used in Europe.*											
3	3.5		*Liaise with a pig nutritionist to formulate a diet density which*											
4	4.0		*will satisfy the published daily intakes (cf Close & Cole*											
5	4.5		*'Nutrition of Sows and Boars', NUP 2000). Total litter weight*											
6	5.0		*at weaning can be on some farms over 30% higher than for*											
7	5.5		*average herds.*											
8	6.0													
9	6.5													
10	7.0													

Gilt<10 piglets Sow<9 piglets			Gilt 10 piglets Sow 9 piglets			Gilt 11 piglets Sow 10 piglets			Gilt 12 piglets Sow 11 piglets			Gilt 13 piglets Sow 12 piglets		
Day	Kg	Fed	Day	Kg	Fed	Day	Kg	Fed	Day	Kg	Fed	Day	Kg	Fed
11	7.0		11	7.5		11	7.5		11	7.5		11	7.5	
12	7.0		12	7.5		12	8.0		12	8.0		12	8.0	
13	7.5		13	8.0		13	8.5		13	8.5		13	8.5	
14	7.5		14	8.0		14	8.5		14	9.0		14	9.0	
15	8.0		15	8.5		15	9.0		15	9.5		15	9.5	
16	8.0		16	8.5		16	9.0		16	9.5		16	10.0	
17	8.5		17	9.0		17	9.5		17	10.0		17	10.5	
18	8.5		18	9.0		18	9.5		18	10.0		18	10.5	
19	9.0		19	9.5		19	10.0		19	10.5		19	11.0	
20	9.0		20	9.5		20	10.0		20	10.5		20	11.0	
21	9.5		21	10.0		21	10.5		21	11.0		21	11.5	
22	9.5		22	10.0		22	10.5		22	11.0		22	11.5	
23	9.5		23	10.0		23	10.5		23	11.0		23	11.5	
24	9.5		24	10.0		24	10.5		24	11.0		24	11.5	
25	9.5		25	10.0		25	10.5		25	11.0		25	11.5	
26	9.5		26	10.0		26	10.5		26	11.0		26	11.5	
27	9.5		27	10.0		27	10.5		27	11.0		27	11.5	

Developed by the UK Meat and Livestock Commission – Stotfold Pig Development Unit

Notes on how to use the feed scale

(1)   Assess piglet and sow condition on Day 10

(2)   Select appropriate scale consistent with piglet number and rearing ability of sow (e.g. a highly productive sow with 10 piglets may require the feed scale for a sow with 11 piglets)

(3)   Where deviations from the scale are appropriate (either up or down) record the amounts consumed in the 'Fed' column

(4)   Cross off the days in the 'Day' column as lactation progresses – allowing relief stock persons to refer to and maintain correct feed intake levels

(5)   Record alterations to piglet numbers and change to the appropriate scale

(6)   Feed lactating animals at least twice per day

(7)   Two diet feeding system is recommended; the lactating sow requiring higher energy and lysine levels than the pregnant sow

(8)   Ensure an adequate water supply. Drinkers should flow at least 1.5 litres per minute

(9)   Ensure correct room temperature. As sow feed intake increases, room temperature should reduce from 20° to 16°C. Maintain at 16°C for the last 10 days of lactation

(10)   When day time temperatures are high, feed one third of the daily requirement am and two thirds pm.

THE ECONOMICS OF LATER WEANING

Most people in the world target on 21 day weaning, which in many cases means 23/24 days at - hopefully - an average weaning weight of 6.5 kg. But, of course, *beware of averages*, because it some herds in adhering closely to the 'wean-by-date' policy rather than 'wean-by-weight' there will be at least 15% small pigs to deal with. Sure, WBW is a better policy but it either involves nurse sows to deal with the 15% or so underweights, or some back-fostering, which is said to - and probably does - inflame virus diseases like PMWS.

Thus proposals by the EU to put back the legal weaning date from 21 days to 28 days (with 'remissions' for those with excellent nursery facilities as a 'let-out' clause), needs some careful examination.

Concern

This news has been received with concern by many European pig producers on the grounds of:

1. Lowered productivity per sow. On mathematical grounds the extra 7-day breeding cycle length (say 150-157 days) will lower target litters/sow/year from say 2.43 to 2.33 on 100 sows weaning 11/litter, that is 110 weaners less/year.

2. Costwise about 3½ more farrowing crate spaces will be needed per 100 sows (17%), a cost which is offset a little (as the farrowing house is the most expensive real estate on the pig farm) by 3.3% less gestation places and 4% less nursery accommodation (because of the 100 fewer weaners produced per 100 sows/year).

How do these penalties stack up?

I've done some calculations but you can put your own costings to my assumptions.

REDUCED INCOME

Those 110 fewer weaners per 100 sows per year should translate into about 104 fewer finishing pigs sold per 100 sows/year or 1.04 fewer/sow. If we work on a gross margin of €15 (£10.20, $18.60)/finished pig, then with each sow averaging 22 pigs sold/year, delaying weaning by 7 days *will reduce Gross Margin/Sow year by about 4.9%.*

INCREASED HOUSING COSTS

A 17% crate house extension divided among 100 sows amortised over 15 years at 8% borrowings will cost around €125 (£85, $155) per sow every year. Assuming 15 years of throughput at 22 pigs sold/year, *this is €0.33 (£0.22, $0.41) per finished pig increased housing cost* over the extra housing cost's lifetime.

WILL THIS BE REPAID?

I don't know whether you have seen the excellent prognostic trials the feed firm SCA have completed (Varley, 2002) in advance of any lengthening of weaning age. A brilliant piece of anticipation! They tested the performance and economics of weaning age moving up by 7 days, actually from 23 to 30 days, which are both more realistic figures than the proposed legislation of 21 to 28 days. (This is because many people are weaning at 23 days rather than at 3 weeks exactly and target weaning age will need to be 30 days in practice so as not to step below the 28 day legal minimum in age in any one weaning week).

RESULTS

Gross margin per sow rose by 16% - easily overtopping the calculated mathematical reduction of 4.9% from the lower sow productivity per year.

Incidentally SCA's return on breeding farm capital in this trial rose by 10%.

SCA'S PERFORMANCE IMPROVED

Sows: While pigs produced per sow per year dropped by 0.3 pigs
                   (1.1%)...

As long ago as 35 years, we tried 32 day rather than 21 day weaning on our farm. While the performance of the much stronger weaners was superior, their dams lost condition too heavily and the idea was abandoned. Today - we know far more about retaining condition on later- weaned sows.

Finished pigs: Feed costs fell by 10.6%

Fixed costs were 8.7% lower

Total costs were 9.35% lower

Bearing in mind that, mathematically, each finished pig has to bear a €0.33 (£0.22, $0.41) extra housing cost (or thereabouts, you can put your own housing cost figures into the equation, not mine) together with the reduced income from the fewer pigs sold/sow/year totalling €35.94 (£24.45, $44.50), the benefits from the 9.35% reduction in total costs comes to €152.03 (£103.42, $188.23), a REO of 4.2:1.

Suggested conclusions

• From this trial work the likely lowered output/sow/year was reimbursed fourfold - a payback of 4 months.

• And the likely extra housing cost would be recouped 35 times over a housing depreciation period of 15 years, giving a payback of under 6 months.

• No figures were available for any reduction in mortality or morbidity postweaning (which could well be significant in a PMWS scenario) which would help offset the 'mathematical' reduction in pigs sold due to a week's later weaning across the herd.

• I suggest breeders/feeders should do a farm trial and calculate the deficit/benefit balance for themselves.

Reference

Varley, M. (2002) 'Piglet Feeding' the Next 5 Years'. Procs. J.S.R. Tech. Conf. Nottingham Sept 2002.

Further reading, see 'Defending that sow - a checklist' (p. 214)

Why these good results were achieved and what must be done to avoid the management problems from weaning 8.5 kg monsters in place of 'normal' 6.5 kg weaners!

WEAN-TO-FINISH

If you have followed the significant developments in the American hog industry over the past few years you will have noticed, and I'm sure reflected upon, the trend over there to wean pigs at about 16-18 days and put them in one building until they are 125 kg or so, the normal US slaughter weight. One move only, from wean to finish.

"How do they cope" you may well ask, "with a spatial requirement of only $0.1m^2$ for a weaner, and eventually $0.75m^2$ for the finishers? Over seven times more! Let alone the vastly different LCTs (Lower Critical Temperatures) for both weight extremes?" Just about double in fact!

Well now, after a few fits and starts, they seem to be coping pretty well, as Table 1 suggests.

Table 1. WEAN-TO-FINISH (W-F) AND CONVENTIONAL NURSERY/FINISHING (N/F) COMPARED

	W-F	N/F
Number of sites	52	68
Weight in (kg)	4.3	4.3
Weight out (kg)	122.4	115.6
Pigs sold/year*	396,812	387,595
ADG (g)	650	620
Lwt FCR	2.62	2.66
Capital cost/pig space	$187	$162
Total income/pig sold	$154.22	$153.83
	(£84.73, €124.56)	(£84.52, €124.24)

* Pigs started were 409,085 in each case, 3% and 5.25% mortalities respectively.
Source: Stein, reported in National Hog Farmer (Blueprint) October 1999

Double-stocking

The spatial difference is mitigated by double stocking in a specially designed, but simple, wide-span building with large pens. A successful system split sexes at weaning into 80 per pen in discrete rooms of 480. Up to about 30-32 kg they get 0.375m² space, when every other pen is removed to another building, while the other half occupies the vacated pen next door. In this way no mixing is involved, the departing groups stay in their pen groups thus pen integrity is preserved. The result is that each wean-to-finish room now contains 3 pens of 80 pigs, 240 in all providing 0.75m²/pig.

Temperature

This is provided to just above the little pigs' LCT by large broiler type gas heated hovers, and provided stockmanship is alert and dedicated, the pigs do well – I've seen them.

What about the results to slaughter – the acid test of any new grow/ finish concept? Table 2 gives one of several controlled results which seems representative.

Table 2. THE EFFECT OF DOUBLE-STOCKING ON W-F PERFORMANCE

		SS	*DS*	
To 56 days	Pigs in (kg)	5.1	5.1	
	Pigs out (kg)	28.7	26.9	
	Daily gain (g)	420	390	
	Daily intake (g)	700	650	
	FCR	1.66	1.66	
		SS	*DS–S*	*DS–M*
56 days to sale (113 kg)	Daily gain (g)	850	850	840
	Daily intake (kg)	2.23	2.22	2.22
	FCR	2.61	2.61	2.60

Source: Univ. Nebraska (2000)

Key   SS   -   Single stocked
DS   -   Double stocked
DS–S -   Double stocked; pigs remained behind
DS–M-   Double stocked; pigs moved

Comment In this trial the results were largely not significant. However, another trial (Univ. of Illinois, 2001) showed that DS–M pigs improved FCR by 5.44% over DS–S and SS pigs, which was significant at the $p = <0.01$ level.

Costs

Rather surprisingly the custom-built W-F building is not necessarily cheaper. Conversions are, but they don't give the better performance of the custom-designed structures. There are quite a few cost analyses, which show quite clearly that it is the improved performance which usually overrides any structural cost increases compared to a conventional 3-site (breeding, nursery, finish) system.

Table 3 gives the list of reasons why W-F is gaining in popularity in North America.

Table 3. REASONS WHY US HOG PRODUCERS ARE CONSIDERING W-F

---

1. Forcing people into AIAO, pigs no longer being sent to slaughter on a continuous flow basis
2. Multisite production has provided more space options for a wider range of sow herd sizes.
3. Feeders used in W-F are better designed than current nursery feeders (more up-to-date).
4. Feeding is more accurate, the novelty encouraging more dedication by stockpeople.
5. Water is more available and managed better.
6. The space needs of nursery pigs seem to have been underestimated in the past. *
7. Less 'down time' i.e. restocking lag/growout lag. In large units this is significant.
8. Group integrity has been maintained with no mixing and disruption of social order.
9. Easier to encourage eating, sleeping and elimination zones.
10. Workers are happier; less time in washing, moving and treating pigs.
11. Producers seem to be conscientious and keep more accurate records.
12. Turn-key buildings and equipment are there for units considering expansion.

---

* What the writer has been recommending for years!
Source: Adapted from Paul Rouen, an experienced practising pig veterinarian in Minnesota.

Advice

Considering the concept? You must go and see for yourself. Once you have talked to people who have converted an existing building **and** those who have built anew you will 'jump' 5 years of experience. Looking is learning in this case. Don't do it until you have.

And I've found American W-F farmers very hospitable and open about any errors they have made.

A typical American wean-to-finish building to accomodate weaning 480 pigs/week, at 3 weeks pigs are split-sexed into 80 per pen, 10 pens/room. At around 32kg the pens are split into 40's, every other pen in the room is emptied.

National Hog Farmer January 1999. With acknowledgement to Bob Johnson and Steve Pate, Illinois.

DIETARY FATS

A whole new area in pig nutrition merits inclusion in this section. New ideas about fats to make you think.

Most textbooks the livestock farmer is likely to learn from – those that are couched in his language – cover the effect of fat on carcase quality very adequately. Today it is impossible for the general public not to be aware of the debate on food and human health, and in particular, the influence of fat in food.

Most pig farmers and their families now recognise that dietary fat affects their blood plasma cholesterol, high levels of which aggravate heart disease.

They are aware, too, that cholesterol can be reduced by increasing the intake of polyunsaturated fatty acids, called PUFAs (e.g. in vegetable oils, margarine, etc) and that cholesterol is increased by intake of saturated fats (animal fats, butter etc).

Heart disease is not a problem in pigs as it is to their owners, of course. Even so, most pig producers know that an increase in PUFAs in the diet of pigs can cause soft fat in the carcase, causing quality problems to the processor and retailer.

Nutritionists now have sufficient information to design diets which contain the correct balance of saturated to unsaturated fatty acids so as to provide firm carcase fat of a desirable colour at the best economic price. However, PUFA-rich fats containing the oleic and linoleic *unsaturated* fatty acids are more expensive than 'harder' fats e.g. tallow, with their high levels of *saturated* fatty acids such as palmitic and stearic.

So far so good, but it is at this point that the information on the importance of fatty acids in most pig textbooks – apart from those written for academics and nutritionists – gets a bit thin, at least in my opinion.

Latest findings

Fatty acid research has moved on, and is still doing so – gathering momentum too. Soon, if communication of the ongoing findings gets down to farm level, the producer will understand as much about fatty acid nutrition of pigs as he now does about protein and amino acid nutrition, which, as I move around farms I know is light years better than it was ten years ago.

'Pig-friendly' fats

This happy phrase was coined by Drs Cole & Close in 2000 in their excellent book on breeding pig nutrition. It covers two developments in fatty acid provision for pigs which are quite exciting.

CLA

The first involves a form of the PUFA, linoleic acid, called CLA or Conjugated Linoleic Acid. Conjugated means 'paired' or 'joined' – but we won't go into bewildering fatty acid chemistry (as the textbooks do) which causes the lay reader's eyes to glaze over and go off to read about some other more understandable subject!

CLA occurs naturally in milk, cheese and beef and has three beneficial properties where pigs are concerned.

1.   It encourages improved carcase quality by reducing backfat thickness, increasing lean, better loin marbling and, if the meat is a little pale, darkening loin colour. Discussing this with nutritionists recently, the opinion is that these significant findings may apply to carcases which may be a little fatty (16 mm P_2 plus) and a little bit low in lean (53%). Even so it is a move in the right direction, especially as other benefits are available as follows…

2.   CLA accelerates the development of the pig's immune system, doing this by improving the lymphocyte protection against viruses. Lymphocytes are the white blood cells which carry antibodies to fight the invading pathogens. Anything which 'calls up' those defending 'troops' more rapidly affords more efficient natural protection against virus disease in particular.

3.   CLA also boosts cytokine activity. Cytokines are messenger proteins which the pig's immune system uses to direct the appropriate defending lymphocyte white blood cells towards the specific type of invading organism. To use the military analogy again (at the risk of making the immunologist's hair stand on end!) the invader could be an air strike, an artillery bombardment, or an assault by infantry or tanks. Cytokines marshal the appropriate response – there is no point in attacking tanks with machine guns or an aircraft with a howitzer.

From the pig farmer's point of view, we can sometimes keep the premises 'too clean'. While this is certainly a better strategy than keeping them 'too dirty', pigs kept in highly sanitised surroundings over a period of time are less challenged by pathogens and so have a lower defensive immune barrier. This is fine from the point of view of better growth and food conversion, as the pig can use the food nutrients which would otherwise be required to build a high defensive wall into building flesh or future piglets instead. However, it is *not* fine should a sudden influx of a disease-bearing pathogen occur. So the standard of sanitation should be kept high *and* measures taken to keep out sudden disease ingress.

CLA therefore works best in the *better*, (not the less-clean) surroundings, where performance is high but the immune defences acceptably low.

How CLA works in the breeding animal is less clear but its value in baby pig to slaughter feeds is established. Good pig nutritionists ensure that sufficient CLA is formulated-in to such diets.

DHA AND THE OMEGA-3 FATTY ACIDS

DHA (docosahexanoic acid) is also a PUFA (or long-chain fatty acid). Our pigs today are probably deficient in DHA because it is not found to a sufficient extent in cereals and in the proteins (mostly vegetable) we now have to use in pig diets.

Good sources are fish oils or high oil fish meals, but fish are getting scarcer and fish meal has got caught up (unnecessarily?) in the animal protein scare; so the two together combine to make DHA levels low in both pig and human nutrition.

What does DHA do? In both humans and pigs, deficiency hinders the development of the foetus and the infant. It also has an effect on brain and nervous function – and in elderly humans, worsening eyesight and memory loss. Maybe the cod liver oil capsules I've been taking for the last 10 years help me to write these books! In pigs DHA addition in the form of fish oil has been shown to improve boar fertility, greater sperm viability and in one case a 13% improvement in born-alives. (Thankfully I am not in need of these benefits these days!) Scottish experiments have gone on to show that adding fish oil to mid-pregnancy onwards sow diets increased the amount of their piglets' brain tissue. Important? It seems so as such piglets were faster to suckle and absorbed more colostrum. As a result these piglets were more vigorous, less inclined to be overlain and pre-weaning mortality was reduced by 1.5%.

Overfishing

But how do we lessen the effects of fewer fish being caught in European waters especially (due to years of overfishing) and the prime economic source of DHA getting scarcer and thus more expensive?

Work by Dr Sandra Edwards involved industrially-produced algae strains selected for their high DHA content used for 4 weeks in late pregnancy and/or in lactation. As with fish oil, late pregnancy supplementation tended to improve piglet survival, whilst lactation supplementation improved weaning weight (*Table 1*). Algae are easy enough to 'farm' and harvest under controlled conditions – and they are a natural, organic source.

Dr Edwards comments: "Intriguingly, however, piglets from sows which had received the pregnancy supplement, although they were not heavier at weaning, showed a better ability to adapt to the weaning process, with better post weaning growth and heavier 8 week weight.

It seems that whenever we look carefully for benefits of long chain omega-3 PUFA supplementation of breeding diets we find a response. With early reports at recent conferences of improvements in litter size when diets have been supplemented with omega-3 PUFAs, this story is set to run and run."

Table 1. THE EFFECT OF ALGAL DHA SUPPLEMENTATION OF THE SOW DIET IN PREGNANCY AND/OR LACTATION ON PIGLET PERFORMANCE

					Significance	
Pregnancy DHA supplement	+	+	-	-		
Lactation DHA supplement	+	-	+	-	Preg	Lact
No of deaths before weaning	1.1	1.4	1.8	2.0	(0.12)	NS
Mean piglet weaning weight (kg)	8.5	7.6	8.1	7.6	NS	0.03
Post-weaning week 1 gain (g/pig/day)	291	338	280	260	0.03	NS
Weight at 26 days post weaning (kg)	21.8	20.6	20.4	19.2	0.03	0.05

Reference

Edwards *et al.*, 2003, Proc. Occasional Meeting of the British Society of Animal Science, Nottingham.

A CHECKLIST FOR PARTIAL DESTOCKING

Many pig units have to repopulate at some time. The time comes when the drag on performance or profit overtops the benefits of just trying to soldier on against the build-up of chronic disease (*Table 1*). Protective medication and enforced culling grows and grows into an unsustainable burden. Something has to be done; the bullet has to be bitten.

Table 1. FULL DEPOP:REPOP

All farms reduce performance progressively – whatever steps are taken

DECREASED PERFORMANCE WITH AGE OF UNIT		
Closed unit	*430 sows*	*AI/AO by Weekly Rooms*
Year		*DLG 7-90 kg*
1		703 g/day
2		684
3		689
4		672
5		670
6		661
7		659

NO RECOGNIZABLE PATHOGENS ENTERED THE UNIT

Source: Kingston, 2000

Suggested Action level Where (dressed carcase weight) meat sold per tonne of feed fed has fallen by 33 kg (7-110 kg)

Depop:Repop (a shortened term for "Depopulation then Repopulation") certainly lifts the drag on performance (*Table 2*). But the main problem with Depop:Repop to date is not only the worrying loss of income over the cleanout period, but that the new herd will consist entirely of gilts, and that the productivity and cash flow for the new herd do not peak until the 3rd parity **which will be 18 months to 2 years after repopulation**. The same goes for full natural immunity protection, although the surroundings the new herd finds itself on will be much cleaner as a result of the fresh start.

Table 2. THE BENEFITS OF REPOPULATION

	Pre restock 12 months	Post restock 12 months	3 years average Post restock
Born alives/litter	10.45	11.62	11.49
Pigs weaned per sow per year	20.5	24.3	23.6
Daily gain (g) (to finished kg)	472 (94)	630 (90.5)	597 (87.9)
Mortality weaning to finish %	14.3	4.3	4.4
FCR (from 25 kg)	2.44	1.94	2.07
MTF* (kg)	325	409	383
% food needing medication	19	<1	<1

* saleable meat per tonne of food fed
Source: Extrapolated from Kingston (2000)

Note: the big rise in daily gain is mostly due to the growing pig being able to divert more nutrients into growth, released from having to use them to build its immune defences against a high level of pathogen challenge prior to repopulation.

Is there another way?

Yes, I'm sure there could be, where periodic, rather than emergency/enforced total repopulation is concerned. Partial repopulation (also called partial de-stocking – PD) has been tried now for over 15 years by a leading pig specialist veterinary practice in Yorkshire, England (The Garth Veterinary Group) with considerable success. Elsewhere recently the system has decreased the incidence of circovirus wasting disease on most farms, as well as Glassers Disease and Strep II Meningitis **but has not eliminated these three** – as it can do with EP, PRRS, Swine Dysentery, APP, PRV, Mange and Lice. It is probably because these other endemic diseases have been eliminated that PMWS/PDNS wasting disease mortality has fallen on PD farms – often substantially (e.g. in my experience 11% to under 3% within one month) once these associated diseases have been considerably reduced or removed entirely.

This said - the more diseases there are on the premises, the greater the risk of a disappointing result in at least one of these diseases. It is essential you P.D. only under the guidance of your veterinary surgeon.

What is partial destocking?

The aim is to remove all growing animals off the base unit leaving only the breeding stock. The base unit is later restarted by weaning into its cleaned-down and disinfected nursery and growing accommodation. Successful producers follow **Four Main Principles**. *The first principle* is that adult breeding stock are normally immune to most of the farm's infections and not excreting the harmful organisms to any significant extent. *The second principle* is that it is mostly the older growing pigs which are re-infecting the weaner pigs with those diseases currently endemic on the farm once the younger pigs lose their maternal antibody protection at, say, 6-10 weeks old. This we call 'back-tracking'. *The third principle* is that during removal of the growers the sows are medicated dependent on the disease profile identified by the farm's regular veterinarian when making routine tests. This is where the diseases I have mentioned which *can* be eliminated are hit hardest. *The fourth principle* concerns the gilts present under 11 months of age. Some farms have removed such animals as their veterinarian suspected they may still be excreting some of the infectious organisms, in particular PRRS, EP and APP. To compensate for their removal from the pig rebreeding flow, either old sows are not culled for one cycle or alternatively, high-health gilts may be purchased to top-up the vacant farrowings needed. These gilts are only introduced at the end of the destock/medication period.

Any in-pig gilts already on the farm must be put into another off-site location and not be re-introduced until they are over 11 months old.

Some rules for partial destocking

- Find off-site grower accommodation for the weaned pigs upwards. While renting buildings is an extra cost, it is always much less than the cost of unattended long-term disease.

- Summer is a good time to do this as cattle and sheep winter housing lies empty and unused.

- Consider stopping gilt purchases over the grow-out period as suggested in the Fourth principle above. A 6 month period has been advised.

- Discuss sow medication/vaccination programs with your veterinarian. EP and PRRS vaccination may be a priority these days.

- The best time-scale protocol for your farm may be different to your neighbour's. Consult closely with your veterinarian.

- Clean and disinfect the nursery and grower accommodation *properly*. For example, follow the 'Antec' schedule. I find this the best yet, although a program called PATHOCLEAN, where the protocols and the products are part of the concept, gives excellent results too (*Table 3*).

- Take the opportunity to modify your nursery/grower accommodation while it lies empty. It can be a good time to change to a batch rearing system with the new weaners.

- Consider moving the unweaned piglets – once they reach weaning weight or date – offsite for 3 to 4 weeks as well.

- Re-evaluate your sources of replacement breeding stock by changing to a supply of 'cleaner' breeding animals if you think it may be worthwhile.

- Reconsider your unit security, especially routes of disease entry into the unit – collection and delivery vehicles especially, and loading ramp disciplines.

- Vaccinate for all the diseases with P.D. is targeted to eliminate if vaccines for them exist, e.g. live PRRS vaccine.

- If other units are close-by (<3km) talk to neighbours about a jointly-timed programme.

- Cull old sows 5 months prior to commencing programme.

- As a result, increase your sow inventory by 15% at least.

- Timing and correct dose of medication supervised by your veterinarian (this will cost up to £30 ($54.60, €44.10) per sow).

Take home messages

- Partial destocking is much cheaper and less traumatic than depop:repop.

- It needs discipline and skill.

- Co-operation and planning with a pig specialist veterinarian who knows your disease profile is essential.

- There are some diseases it will not eliminate, others it can or will do.

- Done well it reduces mortality; makes the unit more manageable.

- It is particularly useful where a farm does not have the financial resources to adopt most of Prof Madec's critical hygiene/management principles. But Madec *and* partial destocking make good bedfellows!

- P.D. is a skilled and meticulous procedure. At the end of the day you may decide to go for broke and totally Depop:Repop.

The British-built 'Trobridge' unit makes an excellent and cheap offsite accommodation. Even cheaper are straw-bale walls and a tented roof e.g. 'COn-Tented' design.

The interior of a 'pipehouse' (in Australia a variant of these are known as Ecoshelters). In this case the off-site accommodation is also used to 'stream' pigs (see the appropriate section in this book). Notice how well the previously-affected pigs are doing with a comfortable floor and ample space.

The improvement from a combination of veterinary-guided farm specific destocking advice *and* the right products and protocols used in the cleandown/ disinfection certainly gives good results. (*Table3*).

Table 3. IMPROVEMENTS ON PERFORMANCE FROM PD USING THE PATHOCLEAN PROTOCOL (700 SOW HERD)

	Before PMWS	*During PMWS*	*After Pathoclean*
Post-weaning mortality (%)	5.1	31.8	2.9
Conception rate (%)	81	75	87
Days to 100 kg	173	192	141
Finishers FCR	2.5	3.0	2.28
Grading < 12mm P_2 (%)	81	75	84

The cash benefit per sow in the final column over column 2 after all costs were taken, were €464 (£315, $573)/sow
Source: ACMC Ltd, Yorkshire, England
Full economic calculations to support these figures are given as Appendix 1 in the second reference below.

The bottom line

I calculate that some 83 kg of saleable meat per tonne of feed (MTF) was retrieved from the PMWS period, which pays for a great deal of rented buildings and better cleandown products/extra labour required. What is this in real terms? On a 700 sow herd selling 15,000 finishers/year eating 3000 tonnes of food (approximately) each year that is recovering over 249 tonnes more saleable (deadweight) meat per year. That's worth a lot of money!

Comparisons

PD compared to Depop:Repop and other health-maintaining management strategies will be found in Table 3, page 240.

References

Kingston, N. (2004) 'Health upgrades; disease reduction strategies for finishing herds.' Procs. R.A.C. Pig Conference, Cirencester.
Gadd, J. 'Full Repopulation or Partial Destocking?' (2005) ACMC Technical Update, Issue 35, pps 3-4.

BATCH FARROWING/WEANING CATCHING ON IN EUROPE - THE ECONOMICS EXPLAINED

Batch farrowing - or batch weaning - is a coming 'thing' worldwide. What is not well publicised are the economics of the idea.

Batch farrowing is an important new – well, perhaps 'newish' – development. Quite apart from the advantages to labour flow and pig flow on the unit, see Tables 1 and 2, batch systems have extremely important features from the health viewpoint. Sure, it has had some bad press recently, but planned in well, and operated according to the rules (failure to do either of these has resulted in the grumbles) the advantages do materially outweigh the difficulties and make it very worthwhile. I outline some vital rules to follow later in this article.

Health

Just as we in Europe are beginning (maybe) to overcome the PDNS/PMWS problem we now are encountering other equally puzzling viral diseases. I'm not saying that batch farrowing is necessarily the whole answer to them but batching production, coupled to the 'MADEC 20'[*] rules, assisted by a much more disciplined approach to overall hygiene and disinfection could be the trio which will see these off too?

Why batch production in this context? Because any disease flare-up in any one batch contains the spread to the affected batch and curtails or prevents its spread to other batches. Vet/med costs can be contained to the affected batch, thus saving on this growing area of costs.

Age Segregated Rearing (ASR)

Only a few people are against this concept (mainly on grounds of start-up/conversion costs) where pigs are reared in age and immune status groups by room,

---

[*] 20 basic management, rather than drug-related, measures advised by the French virologist, Prof. Madec.

building or site. Batch production, at present involving 3-weekly 'pulses', fit in best with the sow's natural rhythm as well as the current preference for 21-24 day weaning (except in the USA where 12-day has given way to 16/17 days). And if you prefer to change to later weaning, as I've discussed in this book (pages 209-219), then batching can be in 4-weekly tranches – no problem.

Conventional batch production – weaning one week, breeding over the next, farrowing during the third week, also fits in well to the ASR concept.

Bigger herds; bigger groups

One of the advantages of modern pig production is to have big enough groups from a single source to utilize All-in/All-out (AIAO) to its maximum. Batch farrowing allows this, and also makes better use of housing and labour.

Labour use

In most of the pig farms I visit, the stockpeople are chasing their tails – there isn't enough time available when things go wrong, mend things or the routine is delayed by some unforeseen occurrence. This is as true in the large units as anywhere else, I guess. Those clients of mine who have converted to batch production still have their hours fully occupied, but because of the concentration of similar vital tasks into one period, the hours are worked more efficiently with less movement and unproductive repetition. I have just measured this aspect on two clients' farms who made the change and it was worth nearly 8% lower labour costs viewed against output (*Table 1*).

Table 1. RESULTS* AFTER ESTABLISHING A FULL '3 WEEKLY' BATCH PRODUCTION SYSTEM COMPARED TO PREVIOUS CONTINUOUS FARROWING (24-DAY WEANING)

	2 years previous (5 parities)	*18 months (3 parities) after the batch system was fully operational*	
Average litter size	10.36	11.21	
24-day weaner weight per sow per year (kg)	124	133.6	(+7.7%)
Vet/med costs per kg weaner weight (UK pence)	8.8	7.4	(−16%)
Labour costs per kg weaner weight (UK pence)	27.2	25.1	(−8%)

Summary: Total benefit to batch farrowing 12.4 p/kg
           Less extra cost of farrowing accommodation needed 1.2 p/kg
           Net gain **11.2 p ($0.20, €0.16)/kg**

* *Figures corrected for UK pig price & labour cost movements over the 4½ years involved (it takes 9 months to fully change over)*

So what are the benefits?

Figures are beginning to come through. Kingston reports (2002) as in Table 2, to which I've added some Euro costings.

Table 2. BATCH PRODUCTION: PERFORMANCE OF FEEDING HERDS WEANING WEEKLY AS COMPARED TO WEANING EVERY THIRD WEEK (2 UNITS – 590 SOWS)

	Weekly weaning	*Weaning every third week*	*Improvement*
Daily Liveweight Gain	490g	547g	12%
Feed Conversion Ratio	2.36	2.26	4%
Drug Cost/pig	€3.07	€1.83	48%
Mortality	11.5%	6.6%	43%

Financial advantage €8.48 (£5.77, $10.50)/pig or 8.7% more income

Source: Extrapolated, financially, from Kingston (2002)

These benefits seem to be reflected in performance improvements in other growing finishing herds I've visited. Basically my impression is that you will sell more pigs, your food conversion doesn't change, but your overheads are substantially down. Nationally the opinion is that sow performance is a little better (but not statistically significantly so) and there are opinions that PRRS and PMWS has 'got a lot better' or been mitigated. Maybe other diseases too.

And the drawbacks?

• Varying sow performance can disturb batch 'rythym'.

• Less successful if housing is close together.

• Difficult if a lot of diseases are present. If so, consider P.D. first (see page 231).

The Danes report that 5%-10% more farrowing pens are needed and 10%-15% more weaner space (DS Annual Report – 2001). With increased interest and depreciation of DKK 5 to 6 per weaner, this has to be offset against the increase or gross margin of DKK 20-40 per finisher. This is an average REO* of 3:1.

In Britain (where batch production seems to be catching on quite rapidly) I can put in some more cost:benefit figures in REO terms.

If we assume that housing costs are about 8% of total costs, and that batch farrowing elevates this by 16% (see below) because more farrowing crates are needed in a batch system, at a €95 (£64, $118) all-in pig production cost (as it is

---

*REO = Return to Extra Outlay. The economic benefit set agaimst the extra cost of adopting a system or procedure.

in the UK today) if the current housing cost is taken as 8% of €95/pig = €7.60, then 16% extra cost added to this figure is €1.22, so the REO (Return on Extra Outlay required) is €8.48 (Table 1) ÷ 1.22 or 7.0:1 which is an excellent return, and a good way of using capital.

Rebreeding skills. Where these are questionable (as with less skilled, not so patient or dedicated stockpeople) batch farrowing, in my experience can be a real headache. *All sows must be served within the week following weaning.* Slackness here creates out-of-schedule farrowings leading to a wide diversity of weaning ages/ weights. *Poor oestrus detection* – especially in gilts – leads to this, so read my two essays on Gilt Stimulation (page 169) and Heat Detection (page 171) which will help keep the service schedule on track, and a *gilt pool is essential* to assist this. Secondly *culling needs to be rigidly effected* to remove sows which fall outside the planned service/farrowing groups. Failure to keep on top of these critical factors is expensive and can erode much of the cost benefit quoted above.

Conversion from continuous to the batch production method

Conversion is a skilled procedure needing meticulous planning. The gradual transfer to every sow being on a three week program is lengthy and should take about 34 weeks – thereafter it is plain sailing/routine *providing strict discipline is maintained*. I find this discipline takes about 2 years before all the lessons/ mistakes/omissions come fully home to roost!

That's life! No one is perfect, and the mistakes that occur by forgetting to follow or bending the rules of batch production take a while to absorb, but once learned – are never forgotten or repeated. Despite this, the concept is viable, feasible to employ with *good* stockpeople – as I have shewn from real farm results – and profitable.

Table 3. HOW DOES BATCH FARROWING COMPARE WITH OTHER 'NEW' STRATEGIES?

	Cost	Improved growth	Reduced mortality	Drug use	Approx. pay back time
All-in/all-out	Low	1-7%*	4-6%	29-45%	Variable*
3 week batch weaning	Low	12-15%	40-45%	30-50%	Long
P.D./sow medic.	Fair	25-45%	45-65%	55-70%	9-15 months
Full Depop:repop	High	30-40%	65-85%	70-90%	14-26 months

*Depends on how out-of-date is the farm before AIAO, and current skill in operating a continuous farrowing regime, and quality of AIAO conversion. Source: Kingston, 2004 extrapolation.

Reference

Kingston, N. (2004) 'Health Upgrades: Disease Reduction Strategies for Finishing Herds', Procs. R.A.C. Pig Conf. Cirencester, England, Sept. 2004.

IS STREAMING AFFORDABLE?

Streaming creates a farm environment where a pig, which has been treated for a disease, never returns to its companions, but joins other once-sick but recovered pigs right through to slaughter. I hope this article will encourage others, especially the academics, to explore it further. So far it has been largely sidelined due to the *apparent* difficulties. But are they as bad in practice as they seem to be in theory?

Sceptical

Quite by chance, several years ago I was in conversation with a producer who had been "sold" the idea by his veterinarian. After reflection he said to me: "I can see the logic in it, but it won't work! I can't afford the extra housing needed." I must say my first reaction was to agree with him, but we thought we'd "stream" a section of his unit anyway as at that time I was working on very cheap, very basic housing systems for growing/finishing pigs and could suggest something relatively inexpensive. We'd see (rather sceptically) how the once-sick pigs would do in this affordable housing.

Two more of my clients were persuaded to take part and we've all been surprised at the results.

What is streaming?

Streaming is a new process to try and contain disease-spread on a single-site unit, and also to reduce the effect of disease – which we all get – on healthy pigs. Streaming creates a farm environment where a pig which has been treated for a disease never returns to its companions, but joins an area where there are other once-sick but recovered pigs right through to slaughter.

My first reaction was that it would be impossibly difficult to operate; my second was that the extra accommodation needed would be impossibly expensive.

I have now had two years experience of the system on several farms. In fact neither has been the case, as Tables 1 and 2 suggest.

Table 1. PERFORMANCE OF FULLY STREAMED AND CONVENTIONAL REARED GROWERS

	Conventional	Streamed		
		'Clean'	'Sick'*	Average of clean and sick pigs
ADG	586	721	572	701
FCR	2.86	2.64	2.87	2.67
Liveweight per tonne of feed (kg)†	350	379	349	375
D.C.W. per tonne of feed (kg) †	252	277	248	273
Mortality % (Weaning to slaughter)	8.9%	5.9%	11.1%	6.6%

* Sick = Pigs (in this case 15%) treated for various ailments (scour, coughing, fever/listlessness, etc.) while removed to separate 'hospital pens', were never remixed with the main healthy herd.
† Of all the pigs sold; does not include mortality levels

Table 2. ECONOMETRICS OF STREAMING

		Conventional		Streamed	
	Income per pig	£ 73.52	($133.80, €108.74)	£74.36	($135.34, €109.31)
Less	Feed costs per pig *	£ 40.80		£ 38.76	
	Drug costs/pig	£ 2.06	£48.90	£ 1.89	£47.53
	Housing costs/pig	£ 2.78	($89, €71.88)	£ 3.47	($86.50,
	Labour costs/pig	£ 3.26		£ 3.41	€69.87)
Gross Margin		£ 24.62	($44.81, €36.19)	£ 26.83	($48.83, €39.44)

Value of Streaming in this Trial Income rose by 1.1% and costs fell by 2.8% giving a 9% rise in margin over operating costs.

* *Home mixed, wet-fed. Feed at £175 ($318, €257)/tonne.*
UK prices, Jan - June 1997, a period of reasonable profit.

The streamed pigs were only those clinically affected by disease and needing therapeutic treatment, often in hospital pens which are commonplace now in many countries and mandatory in others, like Britain. Over a period of two years' experimentation the average number of streamed pigs placed in separate accommodation away from the main nursery and feeder herds was 15%; it could be as low as 8.6%, and it never rose above 17.5%. The farms concerned were in the top third category of competence.

Costs

We took advantage of the cheap housing now available to producers – kennels (outdoor and in) for affected nursery pigs, often home-made from straw bales. The feeder pigs, however, were in tented tunnel houses, some with a straw bale base, some the conventional Japanese pipe-house design, pioneered by Mr Ishigami (see reference). Of course, even this cheap accommodation had an on-cost which averaged 25% extra on the housing cost/pig, but it was still only a 1.5% lift in total costs. More labour, too, was involved, but it worked out at another 0.3% on total costs. (Both calculated on a five-year period.)

Benefits

The results are very interesting. Compared to what we do now (return sick pigs once recovered alongside if not among their unaffected companions) the herd run on the streamed principle gave 8.33% more saleable meat per tonne of feed, and mortality from 32 kg to slaughter dropped by a quarter (on a 3% average), to 2.25% .

Specialised treatment of sick pigs which recovered ensured that by slaughter they nearly caught up with the conventionally-reared group in FCR and ADG terms, but they were over 1% lower in dressed carcase weight.

The unaffected pigs, not having previously sick pigs returned within their environment, gave 9.9% more saleable meat/tonne feed compared to the conventionally-reared pigs.

Overall, the herd run on the streamed principle seems to cost up to 2% more to run, but because the feed costs were reduced by 5% the overall cost of production fell by 2.8%. Due to lower mortality and more lean/tonne feed income rose by 1.1%. This resulted in a 9% rise in margin over operating costs, quite a substantial benefit to cash flow.

This suggests that streaming is worth further exploration.

Some thoughts

1.   The performances in Table 1 seem to confirm the suggestions that once-sick pigs (i.e. those needing treatment) are potent shedders of disease organisms during recovery – or even after they have recovered? *Their* immune protection is high due to the challenge and resultant mobilization of antibodies, which is nothing like as solid in their former companions to whom they are eventually returned. The indigenous pigs then have to

mobilize their own defences and even if they develop no clinical disease, the energy/effort needed to boost their immune status places a drain on their (healthy) performance. Table 1 suggests this was a drop of 10% on the three farms studied.

2.   Even though the hospitalised pigs were kept, from then on, in very cheap (but warm and dry) accommodation, this did not seem to affect their eventual performance to slaughter very much. They caught up to a certain extent (more than we expected anyway), in daily growth terms, with the conventionally-managed pigs, and did do so in the more important saleable meat per tonne of food calculation, despite poorer killing-out percent.

Why was this? The separated, once-sick pigs were warm and not over-stocked. We did notice that under the trial conditions, which certainly grabbed the interest of the stockpeople involved in day-to-day management, the once-sick pigs were being looked after more diligently than one might have expected. Two stockmen told me "It's a challenge, and I'd like to see it work." As Dr Vernon Fowler has told us, "good economic performance is not so much a biological phenomenon, but is more influenced by diligent stockmanship".

3.   Why did the streamed sections of all three farms do better than the conventionally-reared sections? We've already discussed the lower disease challenge to the indigenous pigs. But note – in the streamed section the 'clean' pigs were automatically destocked by between 8% and 20% (average 15%) as the diseased pigs left and *were not replaced*. As one often finds, the original pigs were slightly overstocked, and removing 15% up until slaughter could have accounted for some or even much of the benefit ?

4.   Streaming was not done in the farrowing house. But any pigs in a badly scouring farrowing pen (as distinct from mild nutritional scour) were noted and streamed once the move to the nursery at weaning was made.

Such weaners accounted for 8%, 16% and 9% (average 11%) of all pigs streamed, or two thirds of all those eventually streamed, even though 80% of these weaners were thought to be 'more or less recovered' by weaning at an average of 24 days (stockperson's opinion).

What you might do now

1.   Discuss the concept with your veterinarian.

2. Review the concept with your stockpeople. Get their commitment as things like this will not work if they are half-hearted or cynical about it. The streamed pigs need to be kept separate, that's all – they are not in quarantine. But they'll need separate attention, utensils, etc from the clean pigs and a change of overalls before going back into any 'clean' piggeries is wise.

3. Have you spare space*? It needs to be a separate building*, never just a spare pen in the clean pigs' house. A separate *site* is best of all.

4. If you've the space, can you get – or make – the cheap accommodation which makes the extra housing cost bearable?

5. Having checked these out, try a streamed section and see if you can match the results we got. 8.3% more productivity for 2% more costs is a good bargain.

Pipehouses used for Streaming in Japan

See also section on Partial Destocking for an interior view of streamed accommodation.

Reference

Full design details of the Ishigami 'Pipehouse' system of low-cost grower/finishing housing can be found in:

Gadd, J (1993) 'Tunnel Housing of Pigs' Procs. Am. Soc. Agr. Engineers (4th Symposium) pps 1040-1048.

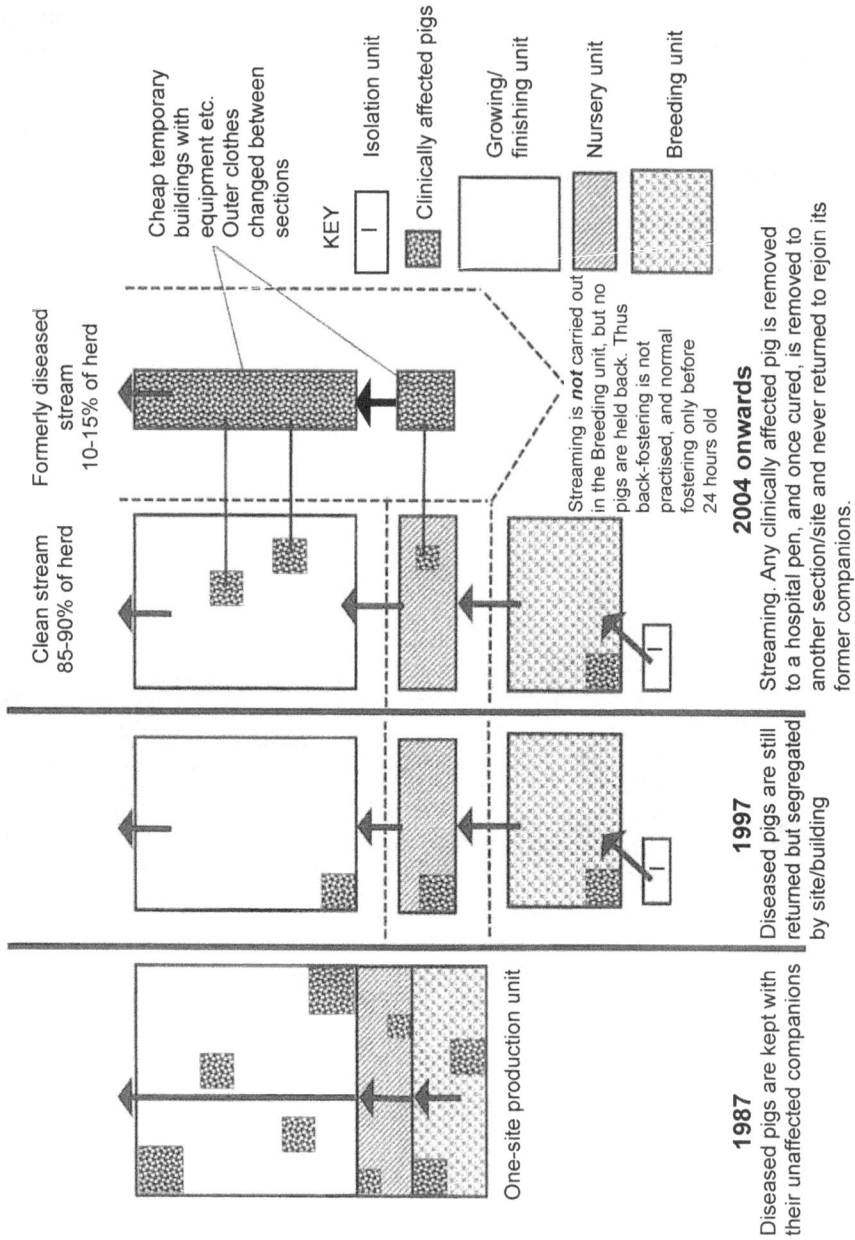

The streaming concept in principle.

CHALLENGE FEEDING/TARGET FORMULATION

It is strange that some innovations in pig production take only months to become universally accepted (slats; sow stalls; wet/dry feeders) while other ground-breaking ideas can take decades to become equally well-adopted (pipeline/soup feeding, All In-All Out; baby piglet nutrition) and only then are adequately covered in textbooks.

One such very promising innovation is Challenge Feeding which leads on to Target Formulation and Farm Specific Diets (FSDs). Even recent nutrition textbooks hardly give the three-cornered, 15 year old concept a passing glance.

Challenge feeding – what is it?

Some 20 years ago in Europe we realised that different pig genotypes (a genotype is what's inside an animal's genetic make-up as distinct from a phenotype which is what the animal looks like) needed rather different nutritional specifications. This resulted in the 'Feeds For The Breeds' concept which soon caught on. 15 years ago over 40% of British pig genotypes were being fed with a customised diet to suit, as closely as possible, breed differences.

Puzzled

After a while we noticed that the same genotypes – obtained from one breeder, not a variety of them – in standard housing but on different farms were giving puzzlingly different performance responses. By as much as 12 kg of saleable meat per tonne of food, which in the money of the time was equivalent to a feed price difference per tonne of 9%.

Any farmer would kill to get a feed that much cheaper for the same performance yield!

Work by American veterinary-oriented pig nutritionists Dritz, Tokach, Stahly and others, then suggested that most of the difference must be the disease status of the different farms' pig populations, particularly the amount of food amino-acid and energy needed to build up a good protective immune barrier where the pathogen or virus challenge was high.

But the problem remained of how to know which farms needed the high immune status diets and which farms didn't – the pigs in each category all looked healthy enough. But those needing the higher immunity barrier just grew slower and converted worse on identical nutritional specs. This was elegantly demonstrated by Dr Stahly's team in 1995 who showed that protein gain/day was 62% more in lean genotype pigs which virtually didn't have to cope with a high disease challenge. With little or no immune demand to satisfy, they would use all the nutrients to grow lean, especially.

So how to determine where the pigs are on the immunity ladder? Quite a problem!

'Test' or 'challenge' feeding

The vets said they could help, but their test would be cumbersome and expensive. So it was decided to 'test' or 'challenge' a representative group (about 50 growing pigs) on each farm twice a year – summer and winter – with what the nutritionists call 'a non-limiting diet' – one where nearly every ingredient is provided a little more generously than the textbooks advise. The pigs help themselves.

These 50 or so pigs are carefully monitored for total growth and for lean growth, using an expensive lean profile saddle-scanner. Environmental conditions, disease status and feed consumed etc. were also recorded and sent to the nutritionist along with the carcase grades, who was then able to construct a lean accretion curve for those pigs on that farm at that time. No, farmers don't necessarily do this work – the feed compounder sends a man to do it, or later, supervise the farmer's own efforts, once familiarity with the procedure kicks in.

From there it is a simple job (for the nutritionist) to design a best-cost diet to meet that particular lean growth curve (Target Formulation). The result is an FSD. (Farm Specific Diet). In American terminology it's a sort of ultra-customized diet – to suit individual buildings if needs be, let alone individual farms, as the performance also takes into account the housing plusses and minuses as well as other variables like overcrowding, ventilation and insulation adequacy, etc.

Results

Not many have been published because the concept was developed commercially, and the pioneer feed companies wanted to keep the good news pretty well under wraps. But I can show you Table 1 which is fairly typical. Anything giving a

21% higher nett profit is worth exploring on your farm, even if the food costs 7% more as in this particular case.

Table 1. EXAMPLES OF CHALLENGE FEEDING RESULTS

	Conventionally-formulated grower/finisher	Challenge tested x 2/year and formula changed every 5 weeks	
Physical Performance			
Deadweight FCR	2.97	2.87	3.36% better
Av. daily lwt gain (g)	786	846	7.63% better
Av. daily saleable carcase gain* (g)	581	660	13.6% better
P2 backfat (mm)	12.1	11.6	
Financial Performance			
Av. cost/tonne of feed	100	107	7% more
Margin over feed cost	100	116	16% more
Nett return	100	121	21% more
REO (Return on Extra Outlay)		3:1	

* *Assuming pigs started with 80% saleable carcase wt at 25 kg.* Source: Clients' records

'Price per tonne' dead-in-the-water?

This brings me on to what could be a major development. *If FSD's catch on universally, then price-per-tonne just doesn't matter any more.* Sure, the FSD food will cost more because the cost of the test and the diet design-work needed (both done by the feed compounder, the first on your farm and the second on their computers) has to be loaded on to their price per tonne quotation.

But if the margin over feed cost is higher (as it nearly always is) then the increased price per tonne doesn't matter, as you'll win in profit terms anyway.

Come to think of it, as most feed selling is still done on price then the expensive feed salesman method of selling becomes largely redundant.

A multiplicity of diets

'Hey!' I hear you say – 'It won't work. No feed company with, say, 500 pig grow-out customers can make 500 different diets!' Sorry, wrong. All the specifications can be made from two, or at most three, diets, delivered to the farm and put into 2 or 3 separate bins. The diets are then *blended* to achieve the correct FSD for the pigs. Either the on-farm mixing producer can do it to computerized instruction from a nutritionist, freelance or company-based, or the feed company can do it from a distance over a land-line at night and deliver the feed ready-mixed. The on-farm methodology just needs organizing and a blender installed – about the size of a large TV set, so it is not expensive.

Sure, this farm blending idea is much easier and so much cheaper with a wet (pipeline feeding) system installed, but I've hammered pig farmers to get into pipeline wet feeding for 35 years and this is just more evidence of the fact that the future of growout feeding almost certainly lies this way.

So I'll finish on a question I cannot at present answer. Why is this futuristic concept taking so long to come universally into the open? You tell me!

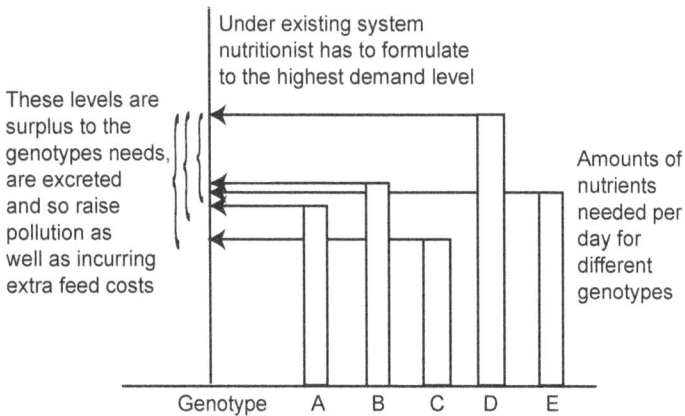

How target formulation reduces pollution.

What we do now . . . restricted range of diets for every circumstance

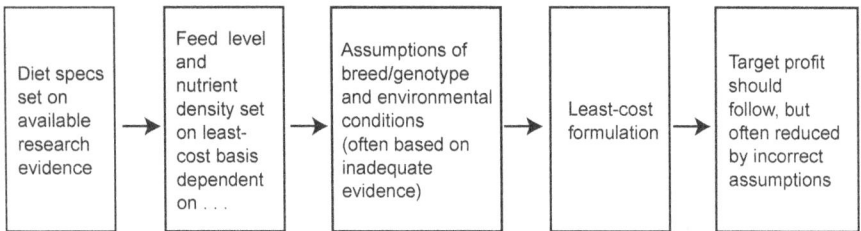

| Diet specs set on available research evidence | → | Feed level and nutrient density set on least-cost basis dependent on . . . | → | Assumptions of breed/genotype and environmental conditions (often based on inadequate evidence) | → | Least-cost formulation | → | Target profit should follow, but often reduced by incorrect assumptions |

How target formulation improves matters . . . specific diets for each farm circumstance

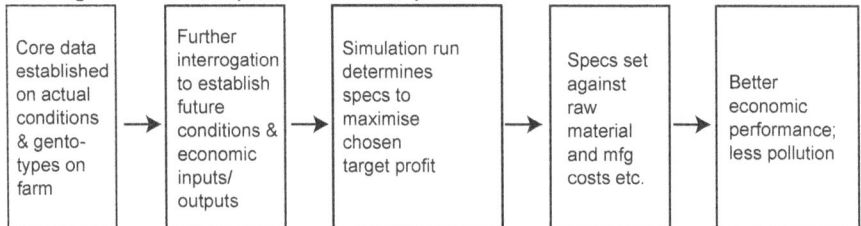

| Core data established on actual conditions & gentotypes on farm | → | Further interrogation to establish future conditions & economic inputs/outputs | → | Simulation run determines specs to maximise chosen target profit | → | Specs set against raw material and mfg costs etc. | → | Better economic performance; less pollution |

Note that with the new method dietary specs are chosen towards the **end** of the assessment build-up, not at the start as at present

MENU FEEDING OF PIGS – UNTYING THE NUTRITIVE KNOT

A LOOK AT SOME DIFFERENT WAYS OF FEEDING THE GROWING PIG

Menu Feeding is one of a variety of new ways the nutritionist can offer food to the growing pig. Unfortunately, worldwide many of these methods are given different names and confusion exists especially among farmers. I subtitled this piece as 'a look back at some different ways of feeding the growing pig' but several of them have yet to catch on, and they may yet do so as automatic feed proportioning gets better and less capital-costly. So let us examine them here.

Menu Feeding of growing pigs is a development of the 1990s recently re-tested in the UK. It is the culmination of 5 different methods of feeding the growing pig; the last two, including Menu Feeding, being restricted (so far) to the nursery stage of growth from weaning at 6 kg to about 25 kg.

To understand how Menu Feeding came about it is essential to understand the progression of the six systems from what most pig producers do now – called Step Feeding. Table 1 summarizes the six procedures.

Table 1. NEW TECHNIQUES IN FEEDING SLAUGHTER PIGS
BUT.... LET'S TALK THE SAME TECHNOLOGY

1. Step Feeding	3 steps	Inaccurate,
	(Steps are Starter-Grower-Finisher)	wasteful, pollution-high.
2. Phase Feeding	5-9 steps	Better, but has physical
		distribution constraints.
3. Multiphase Feeding	20-70 steps	Better still, few distribution
(wet only)		problems, costly conversion.
4.Choice Feeding	2 diets: one high density;	Variable, potentially
Dry or wet, but mainly dry	one low density or just	important. Lowest cost of
	cereals. (Pig chooses diet.)	all.
5. Leapfrog Feeding	2 *balanced* diets, alternated	Moderate cost.
(Choice variant 1)	in sequence, usually in 5-7	
Dry so far	phases/steps. (Pig chooses diet.)	
6. Menu Feeding	As above, but different flavour	Moderate cost, best results
(Choice variant 2)	in each diet. (Pig chooses diet.)	of all so far, but more
Dry so far, wet trials in		comparative trials needed.
progress		

1-4 for all pigs weaning-slaughter. 5 & 6 only used in nurseries so far.

I am being asked many questions about Menu Feeding, because the results are so much better than what we do now, simple Step Feeding (Starter : Grower : Finisher) or even its successor, simple Phase Feeding where there are 5 to 9 steps (eg Starter 1, 2 & 3; Grower 1, 2 & 3; and Finisher 1, 2 & 3). All these are quite easy to program into a modern Computerised Wet Feeding plant.

Let me answer your questions like this:

Likely questions and answers (on the information available) on Menu Feeding of young growing pigs

Q *Isn't this another commercial gimmick?*

A No. The research was pioneered by a reputable feed supplement firm in Britain (Trouw Nutrition) and carried out at the prestigious British Meat & Livestock Commission's (MLC) pig experimental farm, and reported on in November 1995 at a closed conference, which has nevertheless been reported in the British press.

Q *So what's different about it?*

A To understand this, it is necessary to go back to what has happened before...

1. Nutritionists now accept that how the pig is fed in the Acceleration Phase of lean meat deposition (from weaning to about 9 weeks, or 7 kg to 30/35 kg) can markedly shorten time to slaughter (in Europe 92 – 108 kg). The main problem is getting enough nutrients into them at this stage, especially early on.

2. A progression of "PHASE FEEDING" ideas has been tried, in which the pig is given a progressive succession of diets with different nutrient ratios more accurately to suit its changing demand as it grows. This has culminated in multiphase feeding, which is the best yet, but is complicated and expensive to install, as a wet delivery computerised pipeline system is needed (i.e. Big Dutchman).

3. Nutritionists then tried CHOICE FEEDING, allowing the pig to choose its own nutrient ratios from two hoppers in the pen, one containing a high nutrient density diet, one a low one, or just corn/cereals. The two diets were not necessarily balanced in themselves in macronutrients, even if they could be in micronutrients. This had the huge advantages of simplicity and cost.

    Choice Feeding generally works, but there are still unknown factors which need testing before the idea is fully commercially viable.

4. The next stage was to keep the choice idea above but offer two *balanced* diets in the one pen, rather than unbalanced ones and trust

to the pig's instinct to get the proportions right. Called LEAPFROG FEEDING, this allowed 2 phases to be offered at the same time, in the same pen, so that the pig could match his changing nutritional needs more accurately. As one phase feed was withdrawn, usually after 8-10 days, the next phase feed was substituted in the emptied hopper. Thus between 5 or 6 feeds of reducing nutrient density were given through the 8-9 week nursery stage. (*Figure 1*).

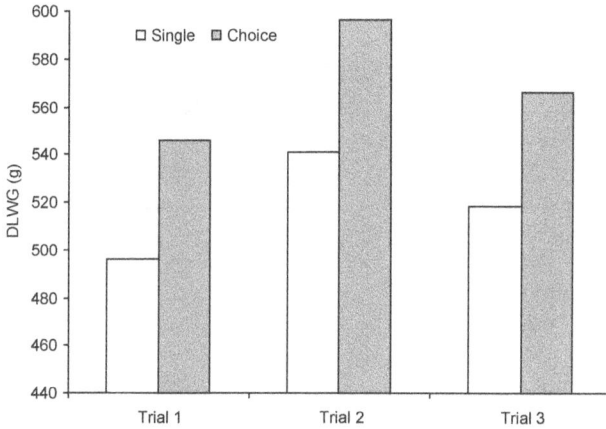

Figure 1. Leapfrog choice: weaning to 30 kg.

The results (over earlier systems as controls) were encouraging (Table 2 overleaf).

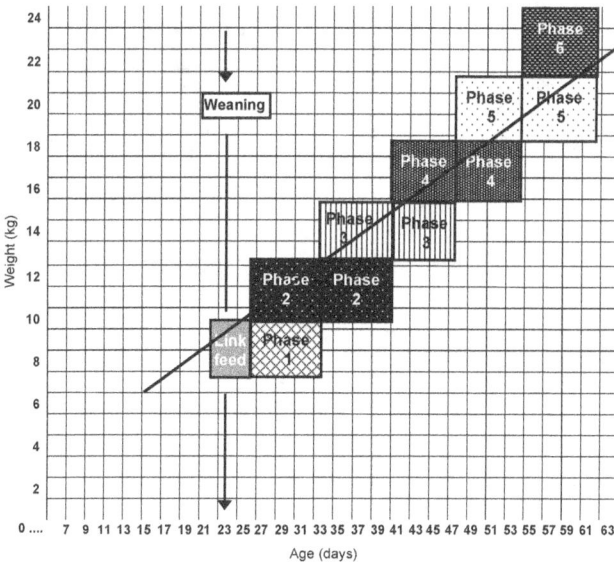

Figure 2. TREATMENT Feed Programme "Menu-Feeding" Trial

Note: For the diets to be available in any one time scale, read the ***upper and the lower*** blocks in the graph

Menu Feeding is a logical progression from this, and can give even more dramatic results if the initial work can be repeated.

Q   *How does it differ?*

A   Quite simply by putting a different flavour into each of the 6 phases (Figure 1). This broke new ground as it assumes, as an omnivore, the pig gets tired of one diet and will eat more if the taste is changed. The results seem to confirm this, (*Tables 2 &3*) and have surprised even the protagonists of the idea.

Table 2. RESULTS: CUMULATIVE DAILY FEED INTAKE (g/PIG/DAY)

Age (days)	Menu	What we do now (Step-feed)	Controls (Phase-fed)	Sig.
31	175	139	149	NS
38	310	247	265	NS
45	440	350	373	p<0.05
52	530	434	460	p<0.05
59	629	507	537	p<0.05

Source: Baker (1995)

Table 3. RESULTS: CUMULATIVE DAILY LIVEWEIGHT GAIN INTAKE (g/PIG/DAY)

Age (days)	Menu	What we do now (Step-feed)	Controls (Phase-fed)	Sig.
31	150	79	99	NS
38	276	205	226	NS
45	382	298	330	p<0.08
52	445	362	383	p<0.05
59	501	407	429	p<0.05

Source: Baker (1995)

Q   *What flavours are used?*

A   It seems that the earlier diets are based on milk-type flavours (i.e. caramel, vanilla) and the later diets on others (ie apple, fish, truffle) *Note*: These were not necessarily the ones used in the trial work.

Q   *Do the flavours or their sequence matter?*

A   Not known at this stage.

Q   *Has the concept been applied to later diet 'phases' i.e. from 30 kg to slaughter?*

A   Not yet, but trials are under consideration.

Q   *Is it expensive to add the flavours?*

A   Based on British figures, the cost of the nursery food increased by about 2% to 4%. As the cost of the amount of nursery food eaten is about 20% of the total feed cost of raising a finished pig (to 88 kg), 2% to 4% of 20% is only another 0.4% to 0.8% on the cost of a finished pig. However, if that pig gets to slaughter only 10 days sooner the feed cost per pig is reduced by 10.5% (at an extra feed cost of +2%), a REO ratio (Return to Extra Outlay Ratio) of 26:1! This is one of the highest REO's known in nutrition, which is why the Europeans are once again taking it seriously. And why any other pig industry should consider doing its own confirmatory trials using existing, commercially-available flavours.

Q   *Why is Leapfrog/Menu feeding such a step forward? Why does the pig seem to respond so positively?*

1. Different pigs housed in one group (and group sizes are getting larger) differ quite widely in their genetic competence to assimilate and convert nutrients. It allows access to better quality feeds for longer periods to those pigs in the pen who need them so as to reach their high genetic potential, and not hold them back.

2. Conversely it allows those piglets with digestive systems which are slower to deal with nutrient changes to remain on lower density diets longer, and thus not cause a digestive stall-out and thus growth interference.

3. The stress caused by dietary change is eliminated, because the pig will always have available a diet to which it is accustomed.

4. Varying the taste seems to encourage appetite.

So .... The better pigs do better; and the slower pigs do better too; they are less-stressed digestively, and may be more contented.

Q   *Final question - what are the snags?*

A   Some things yet to be finalised are:

•   Have we got the right (i.e. the best) flavours from the pig's point of view?

•   Used in the right sequence?

- Does either of these matter?

- What is the best flavour *intensity*?

- Will the pig (as in simple Choice Feeding) choose the right amount of the phased diets if he prefers one flavour to another? Remember he has the choice of two balanced diets, each flavoured differently.

- The complexity of 'rotating' the 6 phases in the nursery hoppers. Difficult (and easy to get wrong on the busy farm), but not impossible, with dry feed delivery systems, but child's play to a Computerised Wet Feeding (CWF) System.

Sure, there are questions to be addressed, but the work so far suggests the prize is so great, and that the exploration work has gone far enough to suggest that any farmer or feed firm/supplement firm should be doing their own trials to build on, test out, and maybe improve on the idea.

We are still in the 'golden age' of pig production, with much to be accomplished.

Reference

Baker, M. (1995). 'Exploiting Growth Potential to Reduce Costs of Pig Production'. Procs. Forum Feeds Technical Conf., Stapleford Park, Oct. 1995.

And on this optimistic note for the future, time to bring this book to a close.

I hope some of the things I've written about have made you think, given you a few econometric benchmarks to use, and most important of all, helped make you a lot more money out of this difficult, ever-changing, price-volatile, but fascinating business of raising pigs for profit.

Good luck anyway.

John Gadd
Shaftesbury, Dorset, England
August 2005

INDEX

www.ingramcontent.com/pod-product-compliance
Lightning Source LLC
Chambersburg PA
CBHW061240220326
41599CB00028B/5486